Centrifugal Pumps:
Design and
Application

Centrifugal Pumps: Design and Application

Editor

Neeraj Mahur

scitus
academics

Centrifugal Pumps: Design and Application

Edited by **Neeraj Mahur**

Printed in 2017

ISBN: 978-1-68117-341-2

Library of Congress Control Number: 2015939253

© 2016 by
SCITUS Academics LLC,
616, Corporate Way, Suite 2, 4766,
Valley Cottage, NY 10989

www.scitusacademics.com

Contents

Preface

Centrifugal pumps perform many important functions to control the built environment. The physics and basic mechanics of pumps have not changed substantially in the last century. However, the state of the art in the application of pumps has improved dramatically in recent years. Even so, pumps are still often not well applied, and become common targets in retrocommissioning projects where field assessment and testing can reveal significant energy savings potential from optimizing pump performance. Typically, retrocommissioning finds that pump flow rates do not match their design intent and that reducing flow rates to match load requirements or eliminating unnecessary pressure drops can save energy.

Editor

The Application of the Ground Source and Air-to-water Heat Pumps in Cold Climate Areas

Kaspar Tennokese, Teet-Andrus Kõiv,
and Alo Mikola, Villu Vares

Department of Environmental Engineering, Tallinn University of Technology, Tallinn, Estonia

ABSTRACT

This article gives an overview of using the ground source heat pump (GSHP) and air-to-water heat pump (A&WHP) in cold climate areas for heating and for domestic hot water production of buildings. Computer simulation and analysis were carried out for a typical detached house, with 200 m² of living area, the heat demand of 9 kW and the average heat demand for DHW production of 1 kW. In heating period the average Coefficient of Performance (COP) of the A&WHP is considerably lower than COP of the GSHP.

INTRODUCTION

In cold winter areas, minimization of fuel cost for heating of buildings is a very important issue. The efficiency of wood stove heating is about 0.5. This means that it is necessary to burn 6 kWh of wood energy to produce 3 kWh of heat (thermal) energy. The efficiency of gas-boilers is about 0.85, which means it is necessary to burn 3.5 kWh of gas energy to produce 3 kWh of heat (thermal) energy. The efficiency of direct electrical heating is about 1, i.e. to produce 3 kWh of heat (thermal) energy, 3 kWh of electrical energy is needed.

Heat pumps are one of the most energy efficient equipments of heating available nowadays. Heat pumps do not produce heat; they simply move available heat from one level to another. The needed electrical energy is predominantly used to run the compressor.

If the average Coefficient of Performance (COP) of the heat pump is 3 (usually from 2.5 to 4.5), 1 kWh of electrical energy is needed to produce 3 kWh of heat (thermal) energy.

The ease of use of heat pumps is comparable to those of simple electronic household appliances: their work is completely automated and they do not require supervision. Heat pumps are suitable for heating and cooling new as well as old buildings.

In this article two types of heat pumps are treated. They can be classified as:

- Devices that draw heat from the ground (with horizontal ground heat exchangers) and heat the building are called ground source heat pumps GSHP, Figure 1;
- Devices that draw heat from the external air and heat the building using a water heating system are called air-to-water heat pumps A&WHP, Figure 2.
- Comparing the efficiency of GSHP and A&WHP in cold climate areas it is necessary to pay attention to the following aspects:
- A&WHP should be operated at temperatures of ambient air below −20°C or even down to −30°C and using additional heat sources (electrical heaters) might be necessary [1];
- Soil temperatures in the depth of the horizontal ground heat collector decrease to the freezing point (0°C) in winter and the soil around the collector pipes might be frozen [2];

- A rather low floor heating water temperature regime (35/30°C) should be used in order to achieve the best possible COP value of heat pumps [2].

Background

Different types of heat pump systems have been studied in Nordic climate conditions in order to find the most cost-optimal heat pump system. In practice, the heat pump was first introduced in 1939 in Zurich Town Hall. Heat pumps began to spread after the Second World War

(Mainly reversible air-to-air heat pumps were used for cooling in summer). Ground-source heat pumps, in combination with different heat sources, have been tested with different system designs in several countries during the last 25 years. The control system has, by the microprocessor technique, opened new possibilities for operation strategies and makes it possible to design and optimize systems for different applications.

The most common type of the heat pump in Sweden today is the ground source heat pump, which extracts heat from a borehole, the ground or seawater [3]. Direct expansion systems are rare today due to negative experience from the period in the mid 1980s when they were popular. They found that the GSHP systems are typically not designed to cover the maximum heat load. They are sized to cover 55% - 60% of the requested heat power at the dimensioning outdoor temperature (DOT). Then they cover approximately 90% of the annual heat energy demand. The remaining part is covered by a supplementary heat source; typically an electric heater, but also oil burners are used.

The website of IEA Heat Pump Centre [4] includes the typical COP values of water-to-water heat pumps at different temperature graphs of the condenser if the flow temperature on the evaporator side is 5°C. It can be concluded from these data that the higher the required temperature is produced the lower the COP of the heat pump will be. It is most useful to use the heat pump for the floor heating system. In the case of the radiator heating system the average COP is 2.5 at the temperature of 60°C/ 50°C and 3.5 at 43°C/35°C. In the case of floor heating (at 35°C/30°C) the COP is nearly 4.0.

ASHRAE Handbook-HVAC Systems and Equipment [5] states that the heat pump can be economical compared to other heating systems when the heating requirement of the building is covered with water temperatures of 45°C - 50°C. This is due to the fact that the COP of the heat pump is relatively high. In this case domestic hot water (DHW) production needs a separate solution because the temperatures of 45°C - 50°C are not always sufficient for producing DHW.

Abel and Voll [6] consider that the heat pump should be large enough to produce 90% of the required annual thermal energy. They also emphasise that for economic purposes it is recommended that the building should also be insulated when a heat pump is installed in an existing building. As the temperature escaping from the heat pump is lower than that from a fossil fuel boiler, already existing heaters can be used if the heat load decreases. The authors note that the low temperature of the heating system is an advantage if the heating system is based on heat pumps, because it is more efficient at low output temperatures.

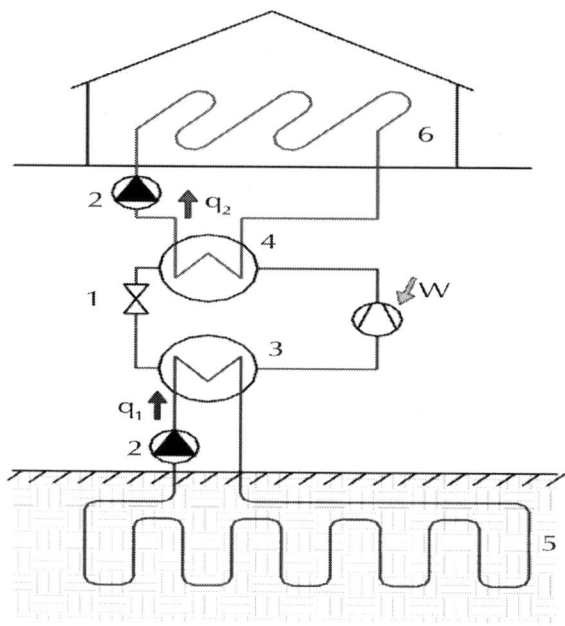

Figure 1: The ground source heat pump: 1—expansion valve; 2—circulating pump; 3—evaporator; 4—condenser; 5—ground circuit; 6—heating circuit.

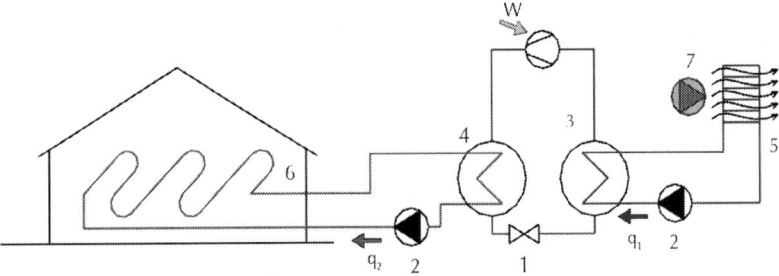

Figure 2: The air-to-water heat pump 1—expansion valve; 2—circulating pump; 3—evaporator; 4—condenser; 5—air-liquid heat exchanger; 7—ventilator.

Penjam [7] has studied heat production with GSHP for heating the building and producing DHW and has compared pollutants emitted into the atmosphere by heat pumps with other more common sources of energy. Dimensioning of heat pumps driven by electricity has been covered in Hayton's [8] research.

Diao et al. [9] preferred using the vertical ground source heat pump because it takes less space. Also GSHP is much more stable than the air to water heat pump. However, when using GSHP, it must be noted that the borehole annulus should be protected with materials that provide thermal contact between the pipe and ground and protect groundwater from possible contamination. Because the vertical borehole penetrates several geologic strata, local specificity must be considered.

Ozgener and Hepbasli [10] tested two different systems: namely a solar assisted vertical GSHP and horizontal GSHP. Test results show that the circulator wattage for the closed loops of GSHP systems I and II can be categorized as efficient and acceptable systems, respectively.

Ground-source or geothermal heat pumps are highly efficient, renewable energy technology for space heating and cooling [11]. This technology relies on the fact that, at depth, the Earth has a relatively constant temperature, warmer than the air in winter and cooler than the air in summer. The geothermal heat pump can transfer heat stored in the Earth into a building in winter, and transfer heat out of the building in summer.

For the optimum design of the GSHP system, it is necessary to estimate its performance and economic feasibility before the introduction of the system [12]. Most analysis models are inaccurate in their predictions for long periods because they are based on a thermal conduction model using a cylindrical coordinate model or an equivalent diameter model.

Tarnawski et al. [13] conducted computer simulation and analysis of a ground source heat pump system with horizontal ground heat exchangers operating in heating and cooling. They found that in spite of the high electricity rate, the ground source heat pump system is a more beneficial alternative for space heating than the oil furnace and the electric resistance system. Besides, the heat pump technology offers relatively low thermal degradation of the ground environment, lower cost of heating and cooling, higher operating efficiency than electric resistance heating or the air-source heat pump and is environmentally clean, i.e. without greenhouse gas emission, if the electricity is generated from renewable energy resources, e.g. the wind and sun. The use of the cooling mode can provide further benefits, such as shorter investment payback and human thermal comfort in summer.

Cui et al. [14] investigated transient heat conduction around the buried spiral coils, which could be applied in the ground-coupled heat pump systems with the pile foundation as a geothermal heat exchanger. Based on the "solid" cylindrical heat source model, an improved analytical model of the ring-coil heat source is established to better illustrate the heat transfer process of PGHE with spiral coils. Coil type heat exchangers are used because the spiral coil configuration has the advantage of more heat transfer area and a better flow pattern without air chocking in the pipes compared to the serial of parallel U-tubes in the pile.

Li and Lai [15] analyzed the influence of anisotropy of anisotropic soil on the processes of heat transfer by GCHP. Several important conclusions are listed: anisotropy of media affects the heat conduction process only over longer time spans, the differences between the cylindrical surface model and spiral line model are very small. So, both can be used to analyze the heat transfer process.

Thermal response tests were carried out in a Korean laboratory [16]. The tests were made to develop an efficient spiral coil source model and its analytical solution, to consider the 3-dimensional shape effects and radial dimension effects of the spiral coil type ground heat exchanger (GHE) using Green's function method. The results of the analytical solution of the spiral coil source model are in good agreement with the overall behavior.

Energy performance of A&WHP and GSHP are studied in the current paper.

METHOD

The ideal heat pump works according to the reversed Carnot' thermodynamic cycle (see Figure 3) and its Coefficient of Performance (COP$_C$) is simply expressed by absolute temperatures

$$COP_C = T_1 / (T_1 - T_2)$$
(1)

Where T_1—the condensing temperature of the refrigerant, K;

T_2—the evaporation temperature of the refrigerant, K.

The working process of a real steam compressor heat pump might be described in a ph-diagram shown in Figure 4. It differs significantly from the ideal, i.e. from the reversed Carnot cycle. The real heat pump cycle looks like a reversed Rankine cycle equipped in addition with an expansion valve. In the valve (line 1 - 2) the pressure of the saturated liquid working fluid abruptly decreases causing auto-refrigeration and flash evaporation of part of the liquid. In the evaporator (line 2 - 3) the cold mixture of liquid and vapor receives heat q_1 from the low temperature heat source and completely vaporizes. In the compressor (line 3 - 4) the pressure increases and the working fluid converts into superheated vapor. In the condenser the superheated vapor is at first cooled and afterwards converted into liquid (line 4 - 4' - 1). In the condenser heat q_2 is relieved and might be used for space heating.

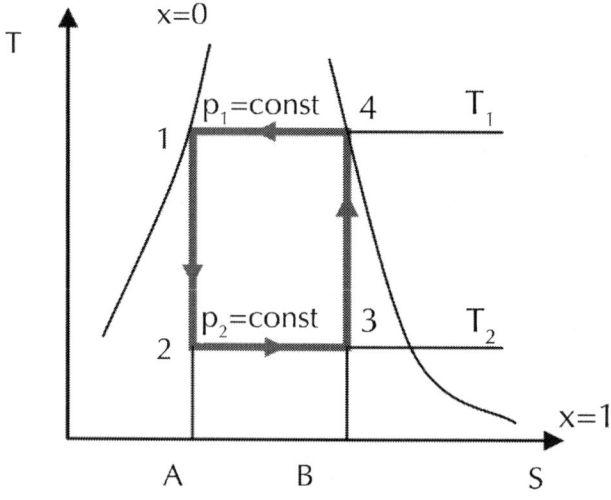

Figure 3: The reversed Carnot' cycle in a Ts-diagram.

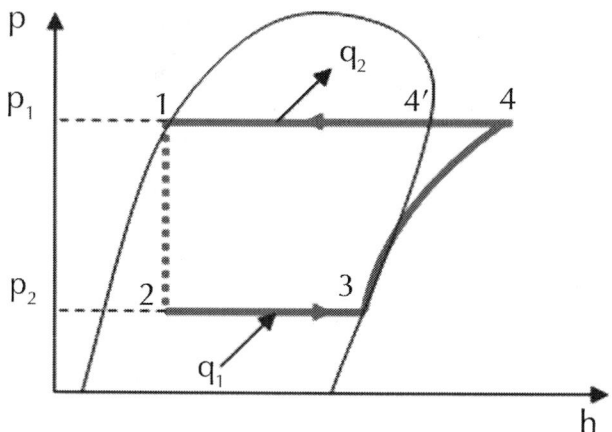

Figure 4: The working process of a real steam compressor heat pump with heat regeneration in a ph-diagram.

Heat pump systems for space heating consume energy not only for driving the compressor but also for other purposes in the system, incl. driving pumps and/or fans and for compensation of heat losses.

The relative efficiency of a real heat pump heating system might be expressed as

$$\varphi = \frac{COP_{HP}}{COP_C}$$

(2)

Where COP_{HP}—the COP value of the heat pump heating system

φ—the relative efficiency of the real heat pumps heating system.

In the case of small-scale heat pumps for heating a detached house, the value of φ might be about 0.5, and in the case of advanced powerful heat pumps for district heating up to 0.7.

In cold climate areas the temperatures of energy sources (soil and ambient air) vary to a large extent during a year (see Figures 5 and 6) and annual (seasonal) electricity consumption and the Coefficient of Performance (COP_{HP}) should be calculated by simulating the heating process of a building).

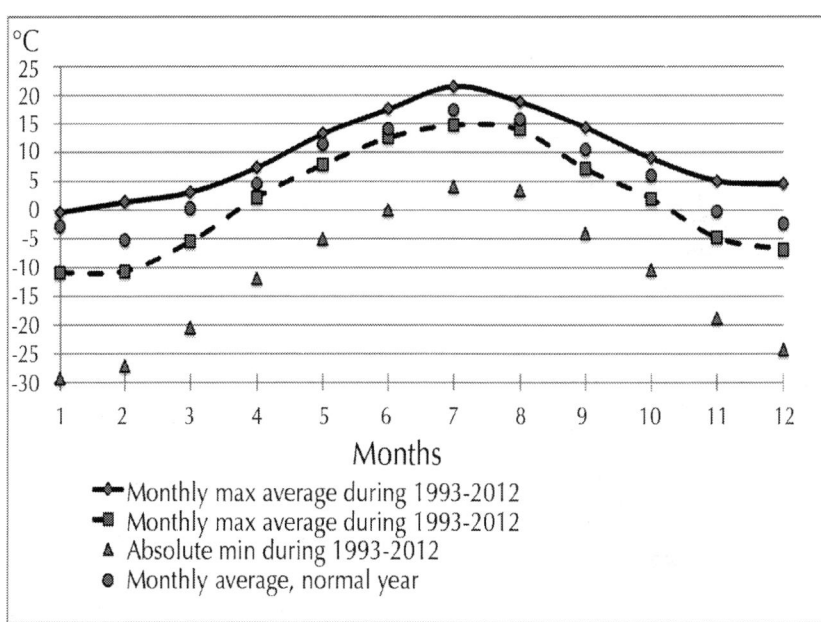

Figure 5: Ambient air temperatures in Estonia (Tallinn, 59° north latitude) during 1993-2012.

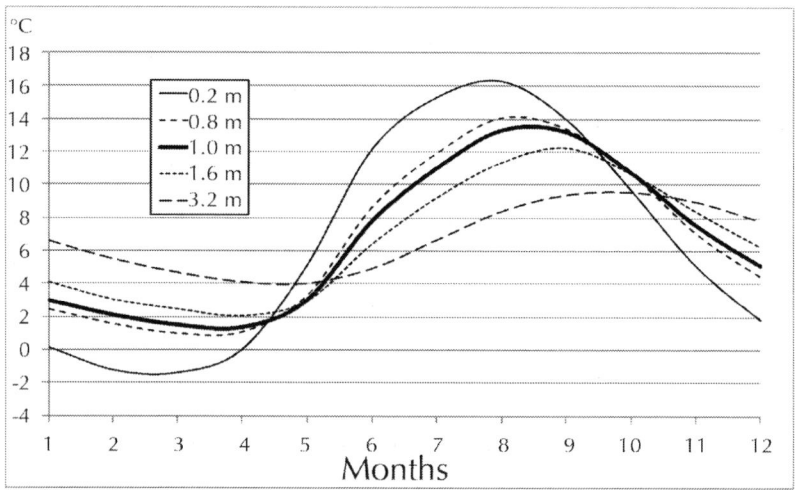

Figure 6: Average soil temperatures in Estonia at the depths of 0.2 - 3.2 m.

Simulation of the heating process allows taking into account the real changes in conditions during the year, such as

- Ambient air temperature profile during a year in case of A&WHP and energy demand for defrost process of the surfaces of evaporation heat exchanger;
- Soil temperature profile during a year in the depth of ground heat collector (1 m in this study) in case of GSHP;
- Design and temperature regime of the water heating system of a building;
- Hot tap water demand;
- Technical parameters of the heat pump.

Long term measurements (about 30 years) at the depths of 0.2, 0.8, 1.6 and 3.2 m have been performed by the Institute of Geology of the Estonian Academy of Sciences. Temperatures at the depth of 1.0 m have been calculated.

RESULTS

The existing building stock in European countries accounts for over 40% of the final energy consumption in the European Union (EU) member states, of which residential use represents 63% of the total energy consumption in the building sector. Improving energy performance in the building sector is therefore a key factor to achieve the EU's climate and energy objectives for 2020, including the reduction of CO_2 emissions and 20% energy saving [17]. Therefore it is very important to evaluate the factors that would increase energy consumption already during the design process. In addition, the European energy performance directive for buildings (Energy Performance Building Directive-EPBD) states that the energy efficiency of buildings has to be calculated in the member states [18].

Description of the Building

The object of modeling is a 2-storey detached house (see standard floor Figure 7), which is situated in Tallinn. The height of floors is 3.0 m and the height of rooms is 2.7 m. The temperature of indoor air is assured by floor heating with water parameters 35/30 and the ventilation air is heated with the water based battery.

To avoid the impact of the usage of the building, the calculations are made with standard usage and by unified calculation methodology. The methodology is specified in local regulations [19]. According to the methodology, the ventilation air flow is 0.42 $l/(s·m^2)$ and the demand for domestic hot water (DHW) heating is 45 $l/(pers.·day)$. The house has no mechanical cooling system. The building has also a mechanical supply and exhaust ventilation system with heat recovery (the efficiency of heat recovery is 0.75; the specific fan power is 2.5 $kW/(m^3·s)$; the minimal achievable leaving air temperature is 5°C; the temperature of the supply air to the rooms is +18°C). The living area of the building is 200 m^2.

The thermal transmittances of the envelope are shown in following table (see Table 1). Pilkington Optitherm S4 type of windows are used, made of triple glass and filled with argon. The thermal transmittance of the frame is 1.4 $W/(m^2·K)$ and the area of the frame is 15% of the whole window. The maximum heat loss of the building is 9 kW.

The model has been made exactly according to the architectural plan of the building. The CAD-file of the building was imported to the program, which guarantees sufficiently exact accomplishment of the model. The plans and views of the building are shown in Figures 7 and

8.

The model of the building has been made using IDA indoor Climate and Energy 4.5 (IDA-ICE) building simulation software [20]. IDA Indoor Climate and Energy (IDA ICE) is a tool for simulating the energy consumption, indoor air quality and thermal comfort of a building. It covers a large range of phenomena, such as the integrated airflow network and thermal models, CO_2 and moisture calculation, and vertical temperature gradients.

For example, wind and buoyancy driven airflows through leaks and openings are taken into account via a fully integrated airflow network model. The model has been made exactly according to the architectural plan of the building. The CAD-file of the building was imported to the program, which guarantees sufficiently exact accomplishment of the model [21].

The current study investigates mainly the influence of different heat pump systems on energy performance. The main investigated systems are the ground source heat pump and the air to water heat pump. The results of the simulations of the air to water heat pump are shown in Figures 9-11. The mean loads and energy consumption of the air-to-water heat pump system are included in Table 2.

Table 1: Initial data of the building model

Part of the thermal envelope	Thermal transmittance, W/(m²·K) #	Unit
External walls	0.17	W/(m²·**K**)
Roof	0.20	W/(m²·**K**)
Slap on ground	0.36	W/(m²·**K**)
Doors	1.0	W/(m²·**K**)
Air leakage rate q50	1.0	m³/(h·m²)
Solar factor g	0.59	-
Heat exchanger efficiency	0.75	-

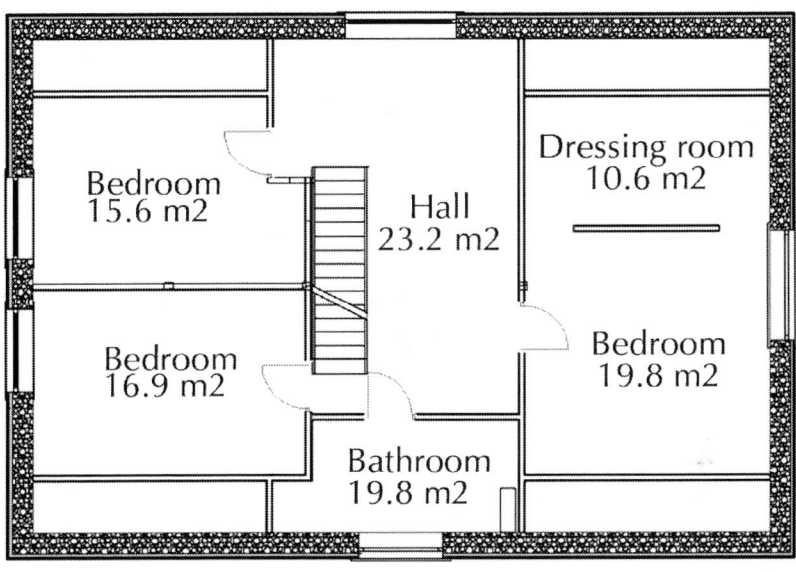

(a)

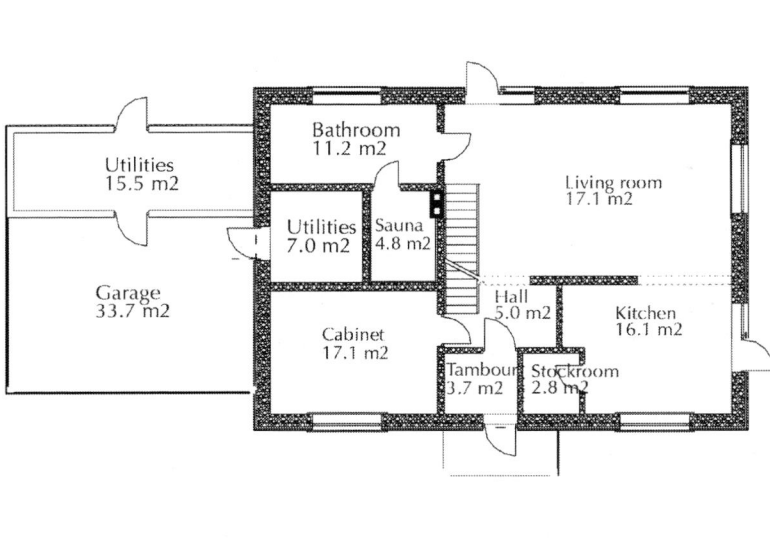

(b)

Figure 7: The ground (left) and first (right) floor of the building.

As shown in Table 2 the annual heat production of the air to water heat pump for heating and ventilation is 16911 kWh, for DHW is 4251 kWh and the mean COP is 2.6. The average COP value of the summer period (May to September) is 2.5 and average COP value of heating period (October to April) is 2.7. In summer period the heat pump produces mainly DHW. The low average COP value of summer period can be explained as the fact that DHW has to be heated to 55°C. At the same time in case of floor heating system the lower parameters (35°C/30°C) are used.

The results of the simulations of the ground source heat pump are shown in Figures 12 and 13. The mean loads and energy consumption of the ground source heat pump system are included in Table 3.

According to Table 3, the annual heat production of the ground source heat pump for heating and ventilation is 17179 kWh, for DHW is 4256 kWh and the mean COP is 2.9. The average COP value of air to water heat pump of the summer period (May to September) is 2.6 and average COP value of heating period (October to April) is 3.2. In heating period the average COP of the air to water heat pump is considerably lower than COP of the ground source heat pump.

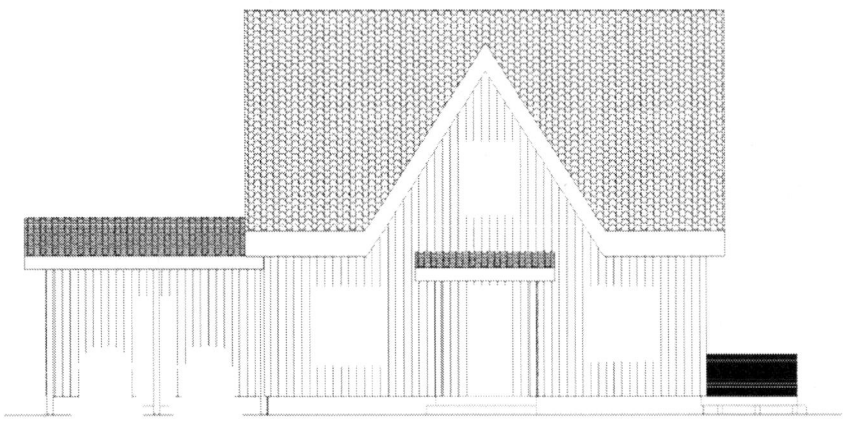

Figure 8: The view of the building.

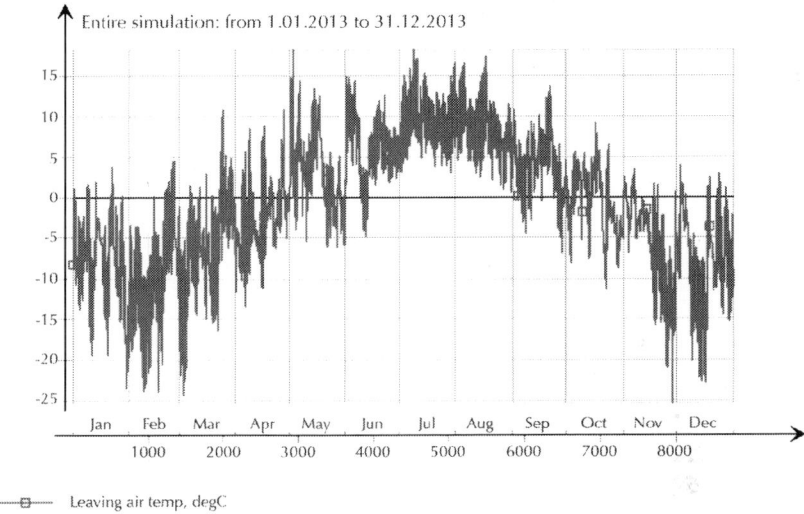

Figure 9: The temperature of the air leaving from the evaporator of the A&WHP.

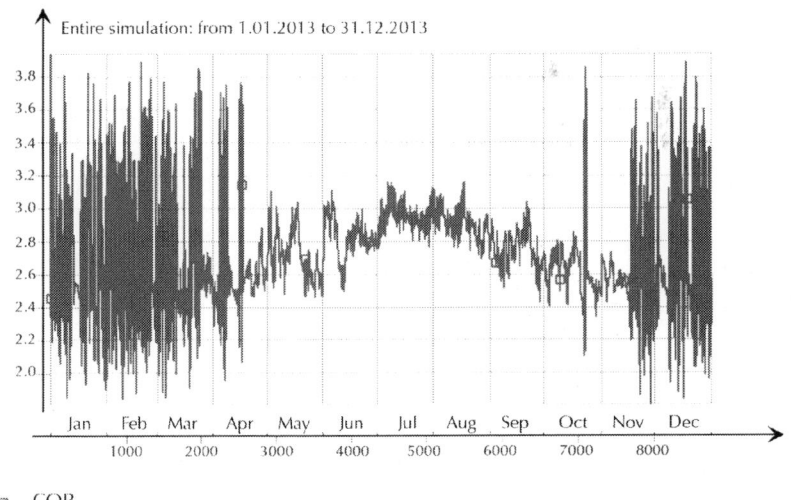

Figure 10: The COP of the A&WHP.

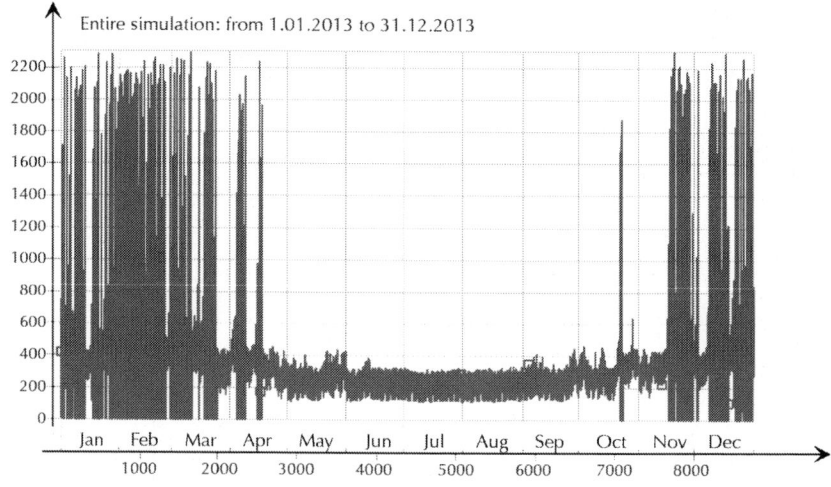

Figure 11: The power of the compressor of A&WHP.

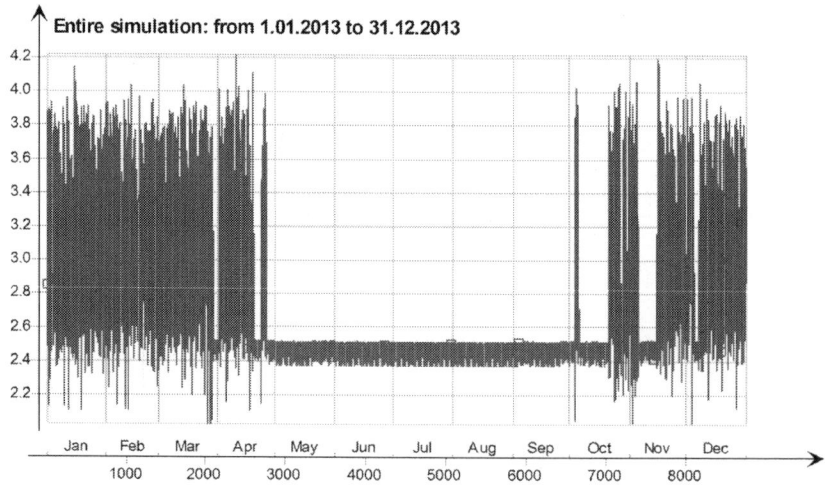

Figure 12: The COP of the GSHP.

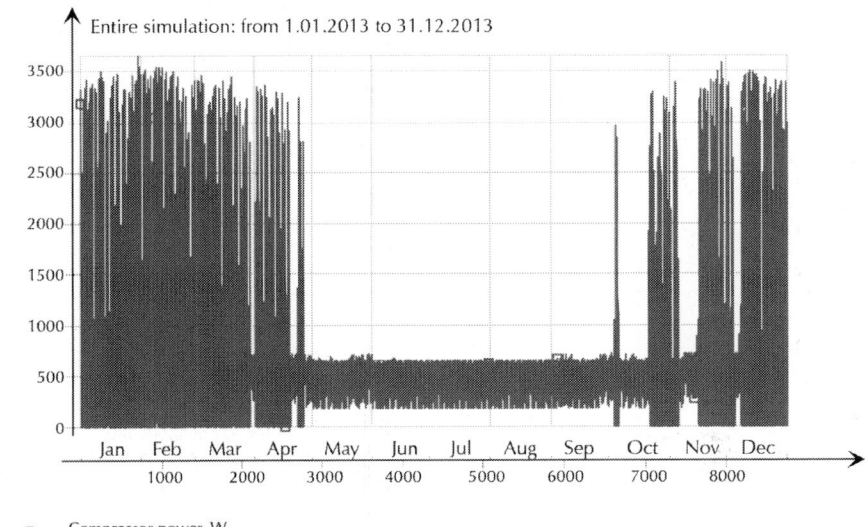

Figure 13: Compressor power of the GSHP.

Table 2: The seasonal energy consumption, COP and mean load of the air-to-water heat pump system

AIR-TO WATER HP	AHU heating coil power, W	Water based heating power to zones, W	Domestic hot water use, W	COP of HP cycle	HP system power, W	Condenser power, W	Evaporator power, W
January	477.6	4218.1	481.7	2.6	2099.4	5255.7	3156.4
February	571.4	4263.6	483.0	2.6	2130.5	5313.6	3183.1
March	359.0	2200.4	485.0	2.6	1253.0	3112.5	1859.5
April	177.7	1072.9	485.8	2.8	703.1	1778.2	1075.0
May	5.8	5.8	486.3	2.4	235.8	566.8	331.0
June	0.0	0.0	486.3	2.5	220.5	549.7	329.2
July	0.0	0.0	486.3	2.6	211.3	550.1	338.8
August	0.0	0.0	486.3	2.6	214.2	549.9	335.7
September	8.1	7.2	486.3	2.4	232.1	562.9	330.8
October	113.0	1409.4	485.8	2.8	788.4	2056.5	1268.2
November	349.2	3412.7	485.2	2.7	1668.1	4288.8	2620.7
December	456.6	4253.3	484.8	2.6	2057.8	5250.4	3192.6
Mean	207.7	1722.8	485.3	2.6	978.1	2470.5	1492.4
Mean*8760 -Wh	1819494.3	15091995.6	4250805.3	-	8568043.3	21641290.0	13073261.0
Min	0.0	0.0	481.7	2.4	211.3	549.7	329.2
Max	571.4	4263.6	486.3	2.8	2130.5	5313.6	3192.6

Table 3: The seasonal energy consumption, COP and mean load of the ground source heat pump system

GROUND SOURCE HP	AHU heating coil power, W	Water based heating power to zones, W	Domestic hot water use, W	COP of HP cycle	HP system power, W	Condenser power, W	Evaporator power, W
January	464.8	4302.8	485.5	3.3	1700.3	5295.2	3594.9
February	558.3	4331.2	485.5	3.3	1735.2	5426.8	3691.6
March	351.9	2241.3	485.4	3.2	1023.6	3112.1	2088.5
April	173.4	1111.3	485.8	3.1	617.7	1827.9	1210.1
May	6.0	6.2	486.3	2.6	222.0	568.3	346.3
June	0.0	0.0	486.3	2.6	215.7	551.5	335.8
July	0.0	0.0	486.3	2.6	215.4	550.8	335.4
August	0.0	0.0	486.3	2.6	215.5	551.1	335.6
September	8.1	9.0	486.3	2.6	221.3	566.1	344.8
October	108.4	1456.5	486.0	3.2	697.6	2082.6	1385.0
November	338.1	3488.4	485.6	3.3	1412.5	4355.7	2943.2
December	443.5	4331.2	485.5	3.3	1703.5	5316.2	3612.7
Mean	202.3	1758.9	485.9	2.9	826.6	2500.6	1674.0
Mean*8760 h	1771750.5	15407534.6	4256483.1	-	7240869.3	21904962.9	14664095.5
Min	0.0	0.0	485.4	2.6	215.4	550.8	335.4
Max	558.3	4331.2	486.3	3.3	1735.2	5426.8	3691.6

CONCLUSIONS

This article gives an overview of using the ground source heat pump (GSHP) and air-to-water heat pump (A&WHP) in cold climate areas for heating and for domestic hot water production of buildings. Computer simulation and analysis were carried out for a typical 2-storey detached house, with 200 m² of living area, the heat demand of 9 kW and the average heat demand for DHW production of 1 kW. The model of the building has been made using IDA indoor Climate and Energy 4.5 (IDA-ICE) building simulation software.

The annual heat production of the air to water heat pump for heating and ventilation is 16,911 kWh, for DHW is 4251 kWh and the mean COP is 2.6. The annual heat production of the ground source heat pump for heating and ventilation is 17,179 kWh, for DHW is 4256 kWh and the mean COP is 2.9. The average COP of the air to water heat pump is 2.5 in summer period and 2.7 in heating period. The average COP of the ground source heat pump is 2.6 in summer period and 3.2 in heating period.

ACKNOWLEDGEMENTS

The research was supported by the Estonian Research Council, with Institutional research funding grant IUT1- 15, and with the project "Development of efficient technologies for air change and ventilation necessary for the increase of energy efficiency of buildings, AR12045", financed by SA Archimedes and by the project "Civil and Environmental Engineering PhD School, DAR9085".

REFERENCES

1.   Ch. Tian and N. Liang, "State of Art of Air-Source Heat Pump for Cold Regions," Proceedings of the 6th International Conference for Enhanced Building Operations, Shenzhen, 6-9 November 2006, Renewable Energy Resources and a Greener Future, Vol. VIII-12-5.

2.   F. Karlsson, "Capacity Control of Residential Heat Pump Heating Systems," Thesis for the Degree of Doctor of Pholosophy, Chalmers University of Technology, Götenborg, 2007.

3.   F. Karlsson, M. Axell and P. Fahlén, "Heat Pump Systems in Sweden—Country Report for IEA HPP Annex 28," 2003. http://www.annex28.net/pdf/Annex28_N28.pdf

4.   IEA Heat Pump Centre, "Heat Pump in Residential and Commercial Buildings," 2013. http://www.heatpumpcentre.org/en/aboutheatpumps/heatpumpsinresidential/Sidor/default.aspx

5.   ASHRAE, "Handbook of HVAC Systems and Equipment," American Society of Refrigerating and Air-Conditioning Engineers, 2008.

6.   E. Abel and H. Voll, "Building Energy Consumption and Indoor Climate (Estonian)," Presshouse, Tallinn, 2010.

7.   V. Penjam, "Heat Pumps COP in the Estonian Context and Environmental Pollution Emissions Compare to Localized Heating System Commonly Used Fuels (Estonian)," Tallinn, 2005.

8.   J. Hayton, "Calculation Procedure for the SAP Appendix Q Process for Electrically Driven Heat Pumps," 2010. http://www.sap-appendixq.org.uk/documents/SAPQ_HP_Calculation_Methodology_19_02_2010.pdf

9.   N. Diao, Q. Li and Z. Fang, "Heat Transfer in Ground Heat Exchanger with Groundwater Advection," International Journal of Thermal Sciences, Vol. 43, No. 12, 2004, pp. 1203-1211. http://dx.doi.org/10.1016/j.ijthermalsci.2004.04.009

10.   O. Ozgener and A. Hepbasli, "Modeling and Performance Evaluation of Ground Source (Geothermal) Heat Pump Systems," Energy and Buildings, Vol. 39, No. 1, 2007, pp. 66-75. http://dx.doi.org/10.1016/j.enbuild.2006.04.019

11.   A. M. Omer, "Ground-Source Heat Pumps Systems and Applications," Renewable and Sustainable Energy Reviews, Vol. 12, No. 2, 2008, pp. 344-371.http://dx.doi.org/10.1016/j.rser.2006.10.003

12.   Y. Nam, R. Ooka and S. Hwang, "Development of a Numerical Model to Predict Heat Exchange Rates for a Ground-Source Heat Pump System," Energy and Buildings, Vol. 40, No. 12, 2008, pp. 2133-2140. http://dx.doi.org/10.1016/j.enbuild.2008.06.004

13. V. R. Tarnawski, W. H. Leong, T. Momose and Y. Hamada, "Analysis of Ground Source Heat Pumps with Horizontal Ground Heat Exchangers for Northern Japan," Renewable Energy, Vol. 34, No. 1, 2009, pp. 127-134.http://dx.doi.org/10.1016/j.renene.2008.03.026

14. P. Cui, X. Li, Y. Man and Z. Fang, "Heat Transfer Analysis of Pile Geothermal Heat Exchangers with Spiral Coils," Applied Energy, Vol. 88, No. 11, 2011, pp. 4113-4119.http://dx.doi.org/10.1016/j.apenergy.2011.03.045

15. M. Li and A. C. K. Lai, "Heat-Source Solution to Heat Conduction in Anisotropic Media with Application to Pile and Borehole Ground Heat Exchangers," Applied Energy, Vol. 96, 2012, pp. 451-458. http://dx.doi.org/10.1016/j.apenergy.2012.02.084

16. S. Park, S.-R. Lee, H. Park, S. Yoon and J. Chung, "Characteristics of an Analytical Solution for a Spiral Coil Type Ground Heat Exchanger," Computers and Geotechnics, Vol. 49, 2013, pp. 18-24. http://dx.doi.org/10.1016/j.compgeo.2012.11.006

17. B. Poel, G. Cruchten, van A. Constantinos and A. C. Balaras, "Energy Performance Assessment of Existing Dwellings," Energy and Buildings, Vol. 39, No. 4, 2007, pp. 393- 403. http://dx.doi.org/10.1016/j.enbuild.2006.08.008

18. EN. Directive 2002/91/EC of the European Parliament and of the Council of 16 December 2002 on the Energy Performance of Buildings, "Official Journal of the European Communities," Vol. 4, No. 1, 2003, pp. L1/65-L1/ 71.

19. Ministry of Economic Affairs and Communications Ordinance No. 63, "Hoonete Energiatõhususe Arvutamise Metoodika. (Methodology for Calculating the Energy Performance of Buildings) (08.10.2012); RT I, 18.10.2012, 1," 2012.

20. Getting Started with IDA Indoor Climate and Energy 4. EQUA Simulation AB, September 2009.

21. T. Kalamees and J. Kurnitski, "Estonian Test Reference Year for Energy Calculations," Proceedings of the Estonian Academy of Science Engineering, Vol. 12, No. 1, 2006, pp. 40-58.

Chapter 2

Study of Secondary Flow Modifications at Impeller Exit of a Centrifugal Compressor

Surendran Anish[1], Nekkanti Sitaram[2], and Heuy Dong Kim[3]

[1]FMTRC, Daejoo Machinery Co. Ltd., Daegu, South Korea
[2]Indian Institute of Technology Madras, Chennai, India
[3]Andong National University, Andong, South Korea

ABSTRACT

A computational study has been conducted to analyze the performance of a centrifugal compressor under various levels of impeller-diffuser interactions. The study has been conducted using a low solidity vaned diffuser (LSVD), a conventional vaned diffuser (VD) and a vaneless diffuser (VLD). The study is carried out using Reynolds-Averaged NavierStokes simulations. A commercial software ANSYS CFX is used for this purpose. The extent of diffuser influence on impeller flow is studied by keeping the diffuser vane leading edge at three different radial locations. Detailed flow analysis inside the impeller passage shows that the strength and location of the wake region at the exit of impeller blade is heavily depended upon the tip leakage flow and the pressure equalization flow. Above design flow rate, the diffuser vane affects only the last twenty percent of the impeller flow. However, below design flow rate, keeping vane closer to the impeller can cause an early stall within the impeller. Small negative incidence angle at the diffuser vane is helpful in order to reduce the losses at the impeller exit.

INTRODUCTION

The nature of impeller exit flow has direct effects on the performance of the diffuser, thence the overall performance of compressor stage. Efforts to understand the behavior of flow inside the impeller started long back. A number of theoretical solutions were developed in the 1950s for isentropic flow through radial and mixed flow impellers. Inoue and Cumpsty [1] measured unsteady variations of velocity and wall static pressure and found that the core of the wake at impeller outlet changed its position, from the shroud to the hub past the suction surface as the flow rate was decreased.

From 1980 onwards, results from Navier-Stokes computations have been attempted to provide additional information about the complex nature of the impeller flow. First reported computations were from Moore and Moore [2]. Later, Denton [3] and Dawes [4] predicted the 3D viscous flow in an industrial compressor accurately. Studies made by Chriss et al. [5], Hirsch et al. [6] and Kang and Hirsch [7] gave a good account of the energy transfer and flow behavior inside the centrifugal impeller. Hirsch et al. and Kang and Hirsch observed that the location of the wake inside impeller results from a balance between the various secondary forces and the tip leakage flow. The balance of the blade surface vortices and the passage vortices depends on the ratio of the streamline curvatures in the blade-to-blade and meridional surfaces, the Rossby number and the ratio of the boundary layer thicknesses of the endwalls and the blade surfaces. Schleer and Abhari [8] described the influence of clearance flow on the flow structure at the impeller exit. The observed secondary flow pattern at the impeller exit behaves similarly to the model proposed by Eckardt [9].

The nature of the secondary flow at the exit of the impeller, as observed by various researchers, is based on the vaneless diffuser. The effect of diffuser vane on impeller exit flow is not much investigated. Shum et al. [10] investigated the upstream potential effect of the diffuser vanes on impeller tip leakage flow.

Present study is aimed to investigate computationally the effects of diffuser vane on impeller exit flow. The diffuser vane leading edge is kept at different radial locations in order to vary the intensity of interaction. Three types of vane diffusers are used; a conventional vaned diffuser (VD), low solidity vaned diffuser (LSVD) and a partial vaned diffuser. For comparison purpose a vaneless diffuser (VLD) is also selected for the study.

COMPUTATIONAL DETAILS AND METHODOLOGY

The centrifugal compressor selected for the study is a low speed centrifugal compressor. Detailed specifications of the compressor are given in Table 1.

As mentioned earlier four types of diffuser configurations are used for the study of impeller-diffuser interaction. The total number of diffuser vanes in VD configuration is 22 and that in LSVD is 11. The solidity (i.e. chord to spacing ratio) for the conventional vaned diffuser is 1.4 whereas for LSVD it is 0.7. In partial vaned diffuser configuration 11 diffuser vanes are fitted each on the shroud wall and on the hub wall in staggered manner. The vanes on the hub and shroud walls are staggered by half vane spacing. The height of the vane in partial vaned diffuser is 6 mm, which is less than the flow passage width. The vane profile is a circular arc for VD, LSVD and PVD. The leading edge shape is made semi elliptical with major axis four times minor axis. Minor axis is the thickness of the vane and it is equal to 4 mm. The trailing edge is also made semi elliptical shape with a major axis of 16 mm and minor axis of 4 mm. All the angles mentioned in Table 1 are measured with reference to the tangential direction.

Numerical simulation of entire impeller blades and diffuser passage require large amount of computational time and computer memory. To avoid this, the computational model used for the simulation of VD and VLD configurations consists of a single diffuser vane and a single impeller blade passage. In PVD and LSVD configurations one diffuser vane passage and two impeller blade passages are modeled. This is done in order to make the area ratio at the interface of rotor and stator close to unity. Figure 1 shows the computational domain of the four above mentioned configurations.

In vaned type configurations the intensity of impeller diffuser interaction is varied by varying the radial gap between impeller blade and diffuser vane (Figure 2). The leading edge of the diffuser vane is kept at two different radial locations; these are,

$$R_3 = r_3/r_2 = 1.05 \,(\text{strong interaction})$$

$$R_3 = r_3/r_2 = 1.15 \,(\text{weak interaction})$$

The diffuser chord length is fixed and is common for all these configurations. At each location the diffuser vane angle is set in such a way that the incidence angle is zero at design flow rate. Impeller tip clearance is set as 2% of the impeller width at outlet.

For steady state analysis the simulation is done at an instantaneous relative position between impeller blade and diffuser vane. Unstructured tetrahedral elements are used for meshing (Figure 3). Fine elements are given around the impeller blade, diffuser vanes and near to hub and shroud wall. Very fine elements are given at the leading edge, trailing edge and at the tip clearance region. Prism layers are attached to the walls. At least six prism layers are kept in the tip clearance region. The geometry modeling and meshing is done using ICEM CFD.

Table 1: Centrifugal compressor details

Speed	3000 rpm
Inducer hub diameter	110 mm
Blade angle at inducer hub	45°
Impeller outlet diameter	393 mm

Number of impeller blades	20
Diffuser outlet diameter	600 mm
Inducer tip diameter	225 mm
Blade angle at inducer tip	29°
Impeller blade outlet angle	90°
Width of diffuser passage	20 mm

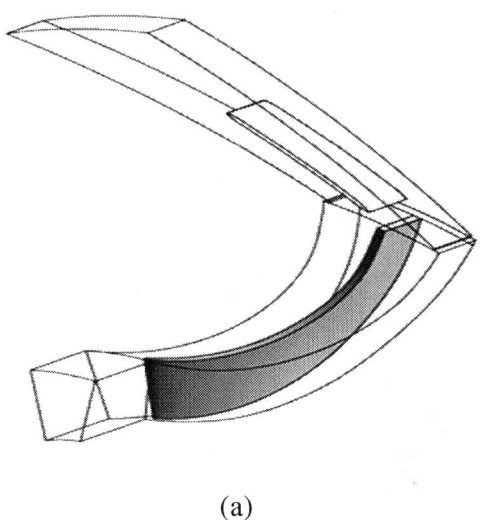

(a)

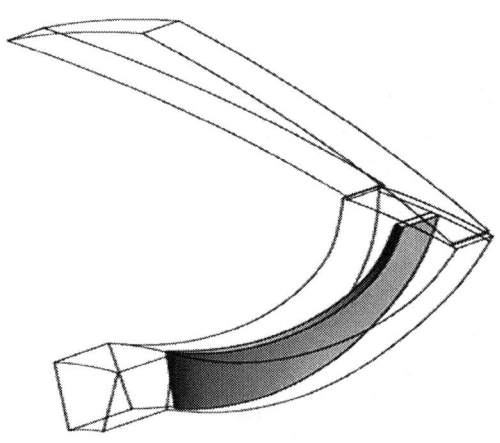

(b)

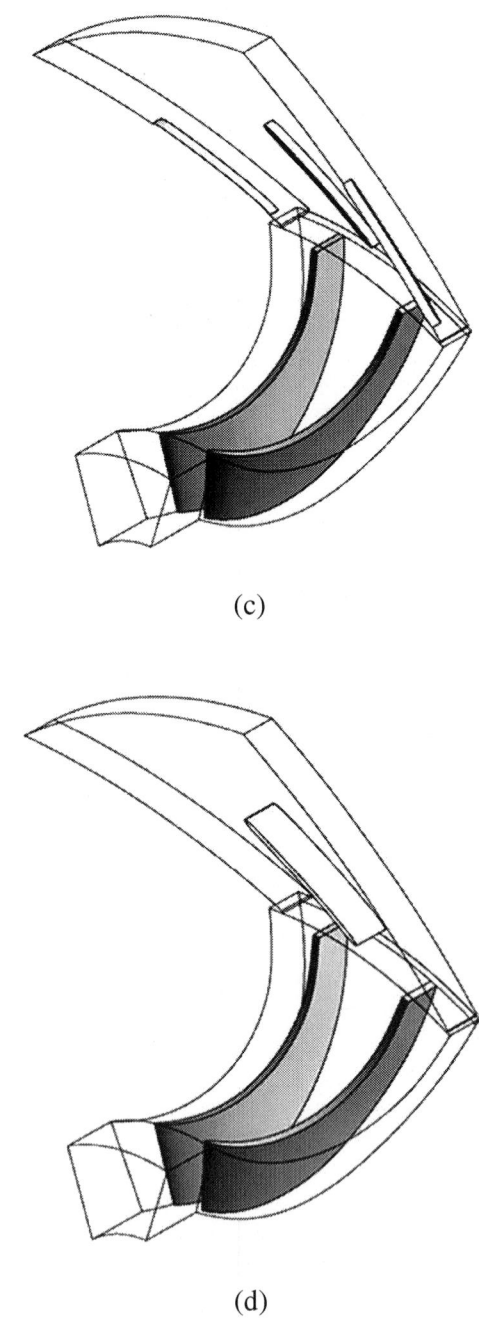

(c)

(d)

Figure 1: Computational model. (a) VD; (b) VLD; (c) LSVD; (d) PVD.

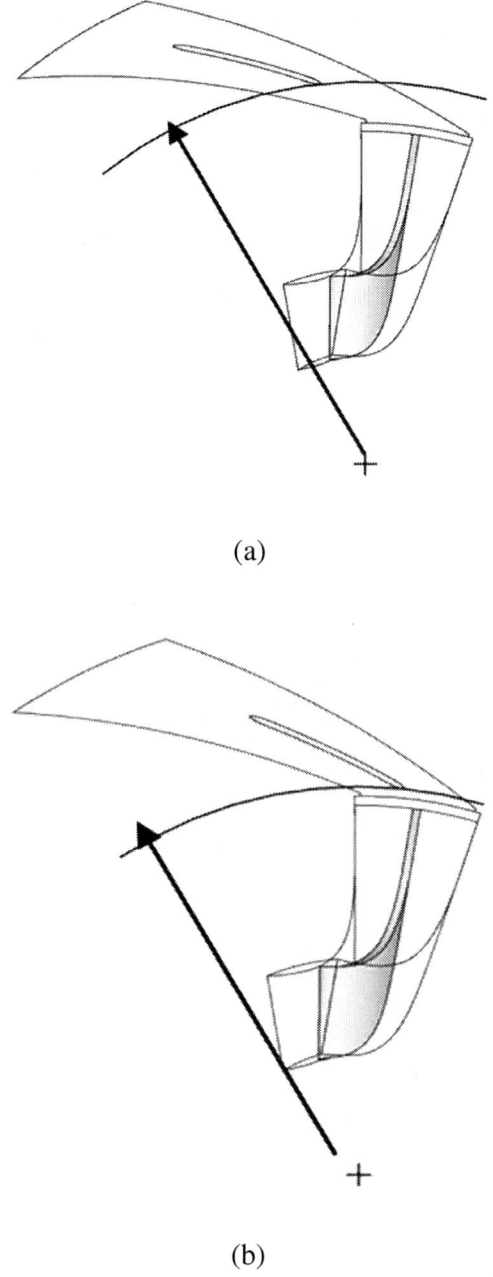

(a)

(b)

Figure 2: Diffuser vane leading edge locations for VD. (a) $R_3 = 1.05$; (b) $R_3 = 1.15$.

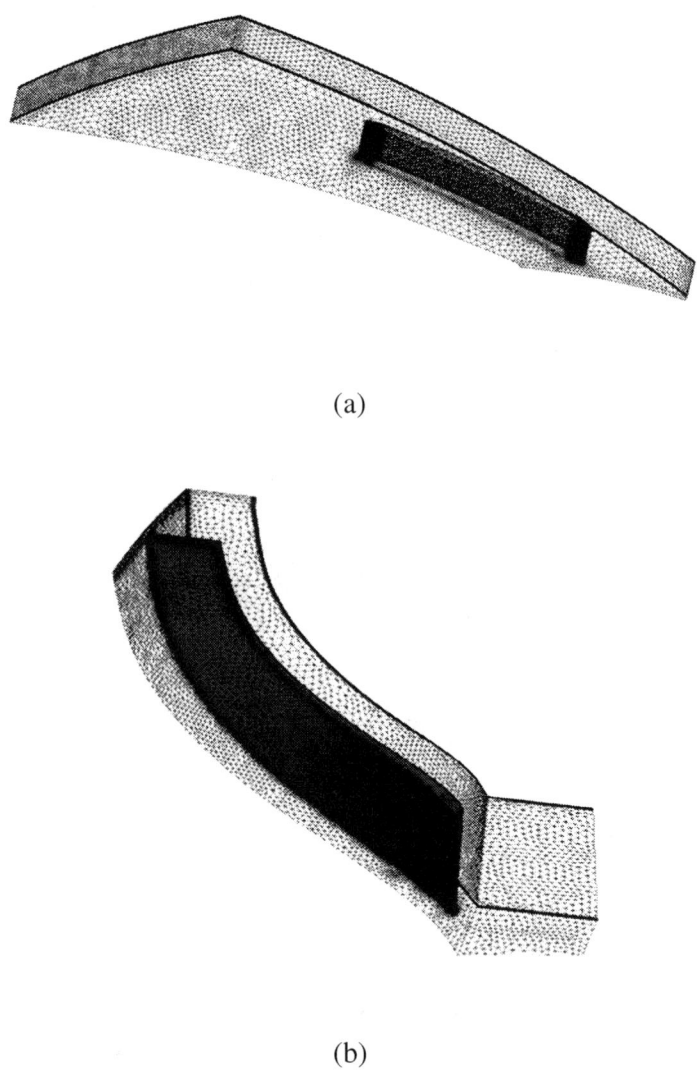

(a)

(b)

Figure 3: Computational mesh. (a) Diffuser; (b) Impeller.

The study is carried out using Reynolds-Averaged Avier-Stokes simulations. A commercial software ANSYS CFX is used for this purpose. Stationary frame uniform total pressure and total temperature is given as boundary condition at the inlet. Specification of uniform inflow direction normal to the inlet plane is quite justified because the inlet boundary is moved away from the impeller.

The fluid used for simulation is air ideal gas. The specified heat transfer model used for the simulation is total energy model. The turbulence model is k and turbulence intensity at inlet is given as 1%. The wall influence on flow of the shroud wall of the impeller domain is specified as counter rotating type, because in a stationary frame of reference the shroud wall does not rotate. At outlet, mass flow rate is specified as boundary condition. As the model contains a rotating (impeller) and a stationary domain (diffuser), suitable periodic interfaces are required between impeller and diffuser. The interface between impeller and diffuser is kept at $R_3 = 1.025$ for all cases and connects both meshes together across the frame change. For the present steady state simulation frozen rotor model is used i.e. the simulation is done at an instantaneous relative position between impeller blade and diffuser vane. This model is robust and uses less computer resources than other interface models. The pitch ratio (i.e. area of impeller domain interface divided by area of diffuser domain interface) is 1.101. As the flow crosses the interface it is scaled to allow this type of geometry to be modeled. This results in an approximation of the inflow to the diffuser passage. To calculate the advection terms in the discrete finite volume equations high resolution scheme is used. ANSYS CFX uses a coupled solver, which solves the hydrodynamic equations as a single system. This solver uses a fully implicit discretization of the governing equations at any given time.

VALIDATIONS

The results obtained from the simulations are compared with the experimental results of Issac [11]. A pre-calibrated three hole probe is used to measure the flow in the diffuser passage. Figure 4 shows variation of mass averaged total $\left(\overline{\psi_o} \right)$ and static pressure $\left(\overline{\psi_s} \right)$ coefficient with respect to radius ratio, R, across the diffuser. Mass averaged value of pressure can be defined as follows,

$$\overline{\overline{\psi}} = \frac{\int_0^s \int_0^b Pcmdxd\theta}{\int_0^s \int_0^b cmdxd\theta}$$

where 0 to s is the circumferential distance between the two consecutive diffuser vanes and b is the diffuser width.

RESULTS AND DISCUSSIONS

Variation of Static Pressure inside Impeller

Variation of total to total (P_t/P_{t1}) and total to static pressure ratio (p/P_{t1}) from the impeller domain inlet to outlet at various interaction levels are shown in Figure 5. The X-axis represents normalized stream-wise distance. This is defined as the distance between computational inlet of impeller and interface between impeller and diffuser along a streamline. The inlet of the impeller domain which is 40 mm away from the leading edge of the impeller blade is represented by "0". The impeller domain outlet which is the interface between impeller domain and the diffuser domain is represented by "1". The pressure values are mass averaged at each streamwise location. Hence this plot provides an overall picture of the diffuser vane effect on the impeller energy addition.

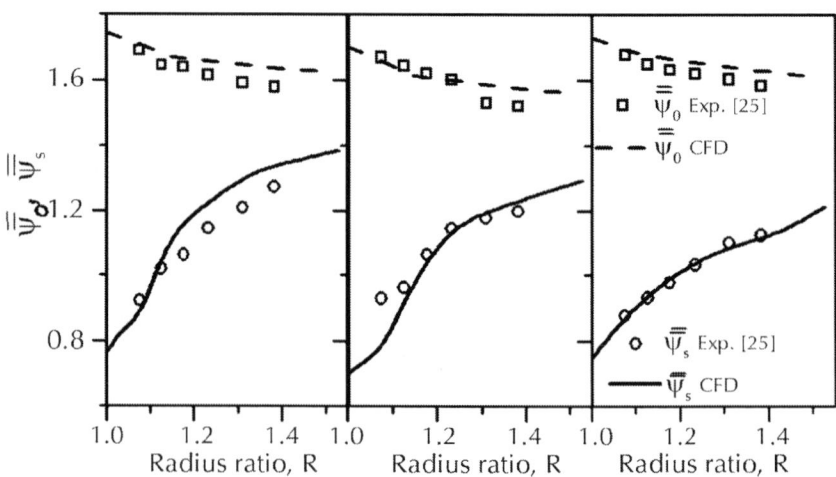

Figure 4: Comparison of experimental and computational results at design mass flow rate. (a) VD; (b) LSVD; (c) VLD.

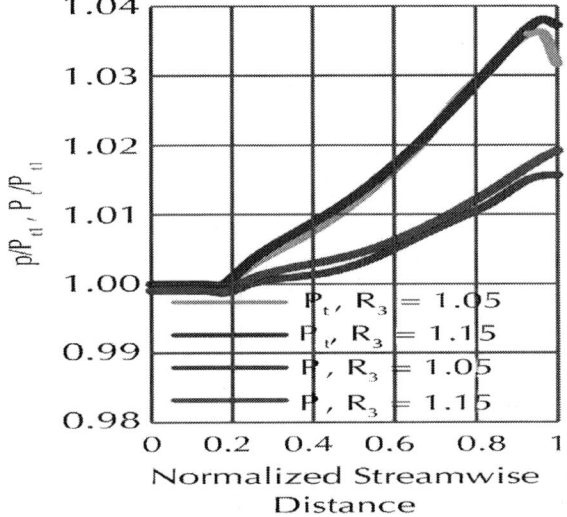

(a)

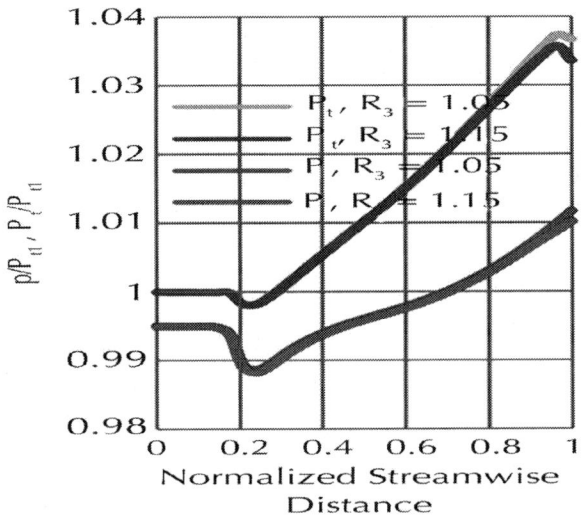

(b)

Figure 5: Variation of total to total (Pt/Pt1) and total to static pressure ratio (p/ P_{t1}) inside the impeller. (a) VD; (b) LSVD.

In the case of VD the variations are shown for above design mass flow rate of 0.965 kg/s (near choke). Differences are observed at the last twenty percent of the impeller domain, which indicates that the effects of interaction are not penetrated deep into the impeller flow at this mass flow rate.

For LSVD the pressure variations inside impeller are shown at below design flow rate (i.e. m = 0.423 kg/s). This flow rate is close to stall margin for LSVD. The static pressure rise is smaller for R_3 = 1.05 compared to R_3 = 1.15 configuration. The pressure ratio curve indicates that the effect of diffuser vane is felt throughout the impeller passage for a closer radial gap. The R_3 = 1.05 configurations may be in the stall region at this mass flow rate. Keeping the diffuser vane away from the impeller can delay the stall inside the impeller. The effect of diffuser vane on the impeller energy addition is very small for PVD configurations (not shown here)

Flow Analysis inside Impeller Passage

Figure 6 shows the relative total pressure on the pressure surface (PS) and suction surface (SS) of impeller blade with at different mass flow rates. These contours are shown for the impeller with a vaneless diffuser at downstream. Relative total pressure represents the flow behavior inside a rotor correctly, as it is the sum of static pressure to the dynamic head based on the relative velocity. Due to the turning of the flow from the axial to radial direction a pressure gradient develops from hub to shroud.

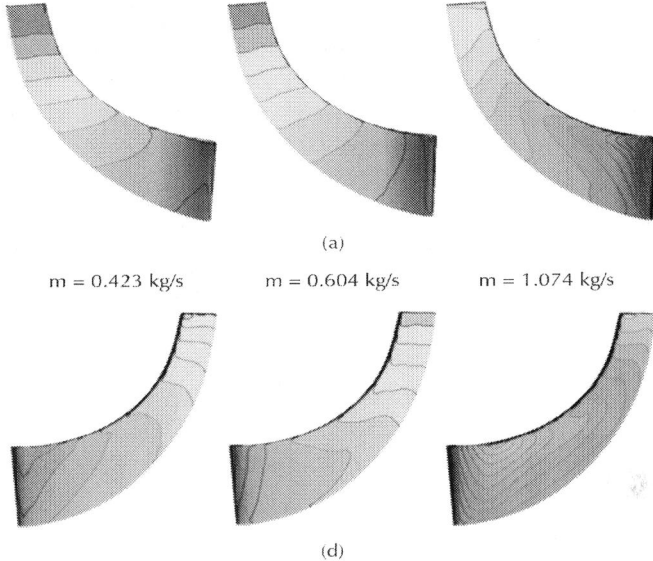

(a)

m = 0.423 kg/s   m = 0.604 kg/s   m = 1.074 kg/s

(d)

Figure 6: Relative total pressure contours on impeller blade (VLD): (a) Pressure surface; and (b) Suction surface.

This pressure gradient in the span-wise direction gradually comes down and becomes uniform at the exit of impeller blade. The pressure gradient in the hub to tip direction increases with the mass flow rate. Low relative total pressure is observed on the suction side near the axial to radial bend. The pressure gradient is higher on the suction surface and maximum gradient is observed on the inducer part. This pressure gradient and the centrifugal force cause the fluid movement from the hub to tip. This movement is observed from the relative velocity vectors plotted on a plane adjacent to the suction surfaces of the impeller blade (Figure 7). The upward movement of fluid is visible near the suction surface where the relative pressure gradients are high.

Kang and Hirsch (1999) pointed out that the relative total pressure gradient in the hub to shroud direction causes the low velocity wake region to move towards the shroud. To validate this, contours of non-dimensional meridional velocity are plotted on a plane normal to the throughflow direction at a streamwise location M = 0.5 (Figure 8). This plot has been shown only for impeller with VLD at downstream. The flow behavior inside the impeller at M = 0.5 is similar for other configurations also at design point.

Near the shroud a low velocity wake region is observed for all the mass flow rates. This low velocity region is spread throughout the shroud wall in the case of m = 0.423. As the mass flow rate increases the core of this wake region move towards the suction surface. This movement is under the influence of Rossby number and it has been reported by Inoue and Cumpsty (1984) and Kang and Hirsch (1999). For the present analysis a radial tipped impeller is used hence the flow at the exit of the impeller is not controlled by the Rossby number alone. Instead the exit flow is mostly under the influence of tip leakage and pressure equalization flow. Figure 9 shows the streamlines inside impeller with vaneless diffuser at m = 0.604. The streamlines show the movement of low energy fluid from the hub to the shroud on both sides of the impeller blade. From the pressure surface (PS) side this low momentum fluid pass through the tip clearance region as a jet flow (note the high velocity of the fluid inside the tip gap). This tip leakage flow interacts with the low momentum fluid on the suction surface (SS) causing the latter to roll down and move towards the exit. At the exit pressure equalization flow from the PS to SS pushes the low momentum fluid back in to the impeller passage. This causes further deceleration of the flow from suction side. The strength and location of the wake region at the exit of impeller blade is heavily depended upon the tip leakage flow and the pressure equalization flow. Shum et al. (2000) observed some variations in the impeller performance with tip leakage flows and also observed that the tip leakage flow is a function of the diffuser vane loading. To understand, how the impeller out flow varies with different types of diffusers, normalized meridional velocity contours are plotted at M = 0.97 (i.e. near trailing edge of the impeller blade).

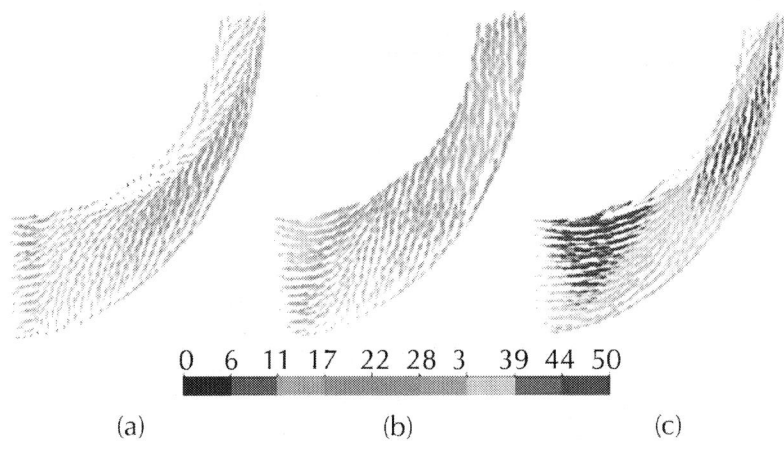

0 6 11 17 22 28 3   39 44 50

(a)   (b)   (c)

Figure 7: Relative velocity vectors near impeller blade suction surfaces (impeller with VLD). (a) m = 0.423 kg/s; (b) m = 0.604 kg/s; (c) m = 1.074 kg/s.

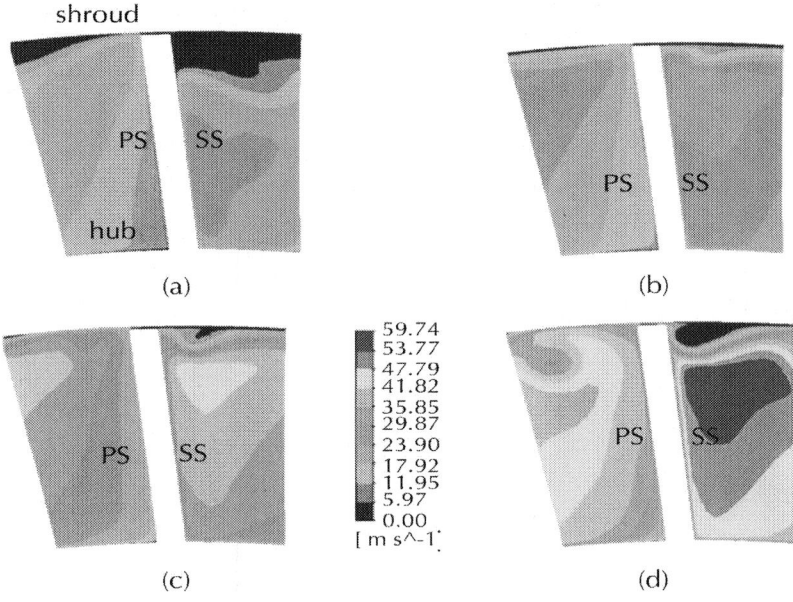

Figure 8: Contours of non-dimensional meridional velocity at M = 0.50, impeller with VLD. (a) m = 0.423 kg/s; (b) m = 0.604 kg/s; (c) m = 0.805 kg/s; (d) m = 1.074 kg/s.

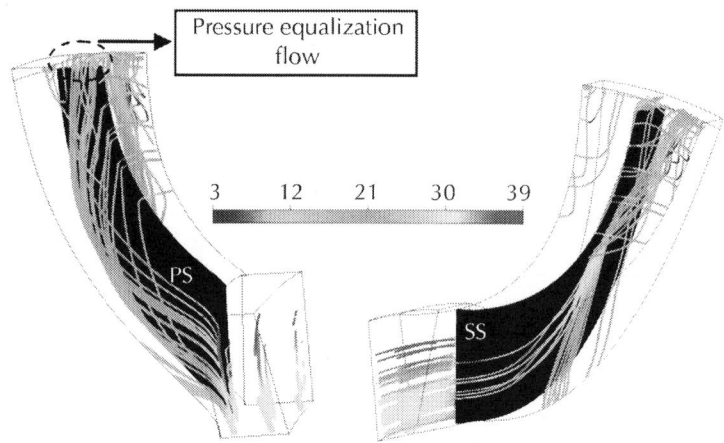

Figure 9: Streamlines showing the pressure equalization flow and tip leakage flow; view from (a) pressure surface and (b) suction surface; m = 0.604 kg/s.

Flow Behavior at Impeller Exit (VLD and VD)

Figure 10 shows the meridional velocity contours at M = 0.97 for impeller with VLD for four mass flow rates. The low energy zone observed at M = 0.5, near the shroud, is strengthened by the tip leakage and pressure equalization flow. As the mass flow rate increases the pressure equalization and tip leakage flow are increases hence at m = 1.074 the wake move closer to the blade suction surface. The jet region, which is observed at the hub suction side corner, moves towards the pressure surface of the impeller as the flow rate increases.

Figure 11 shows the non-dimensional meridional velocity contours at strong interaction level (R_3 = 1.05) for VD. At m = 0.423 large wake region is observed near the blade suction surface. At closer radial gap the back pressure from the diffuser assists the reverse flow near the shroud. At higher mass flow rate the core of the reverse flow region moves away from the suction surface. It is quite interesting to note that at design mass flow rate (m = 0.604) the reverse flow region is much higher than the slightly above design flow (m = 0.805).

When the diffuser vane leading edge is kept at R_3 = 1.15 (Figure 12) the strength of the wake region and reverse flow region reduce are reduced at lower mass flow rates. The nature of the velocity contours is almost similar to that of R_3 = 1.05. Here also, smaller wake region is observed at slightly above design flow rates (i.e. at m = 0.805).

To understand this phenomena in detail, circumferen- tially mass averaged flow angle at M = 1.01 is plotted (Figure 13). At design flow rate, the flow angle is almost constant up to 70 percent of the span. Near the shroud flow becomes highly tangential. The negative flow angle near the shroud indicates the reverse flow at that location. Because of the small flow angle near the shroud, flow does not enter through the leading edge instead it hits on the upper convex surface of the diffuser vane (often referred as positive incidence). The flow then turn around the leading edge and flows along the lower concave surface of the diffuser vane, before it separates (Figure 14(a)). This flow enhances the reverse flow into the impeller passage. At m = 0.805, flow angle near shroud become more radial, hence there is not much reverse flow to the impeller (Figure 14(b)).

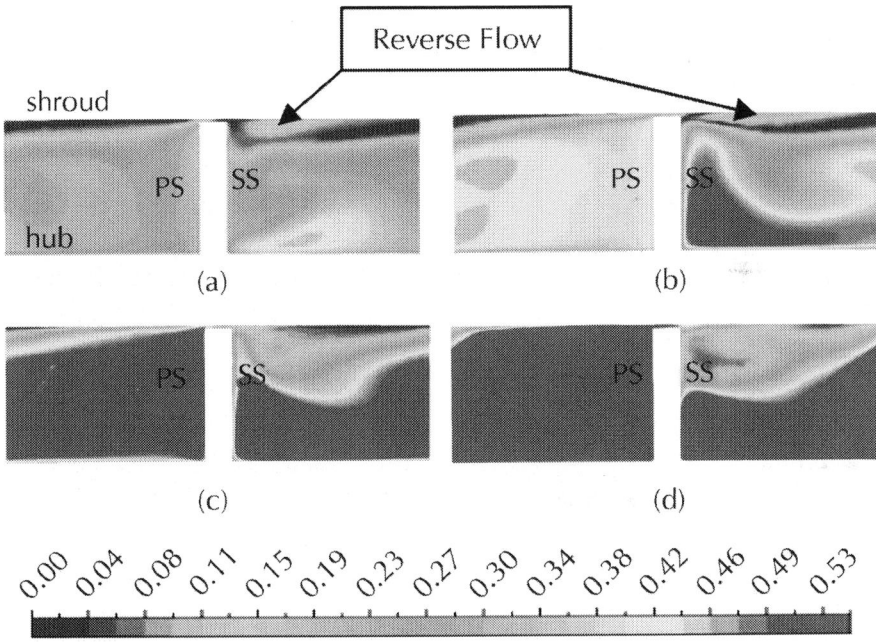

Figure 10: Contours of non-dimensional meridional velocity inside impeller at M = 0.97 (impeller with VLD). (a) m = 0.423 kg/s; (b) m = 0.604 kg/s; (c) m = 0.805 kg/s; (d) m = 1.074 kg/s.

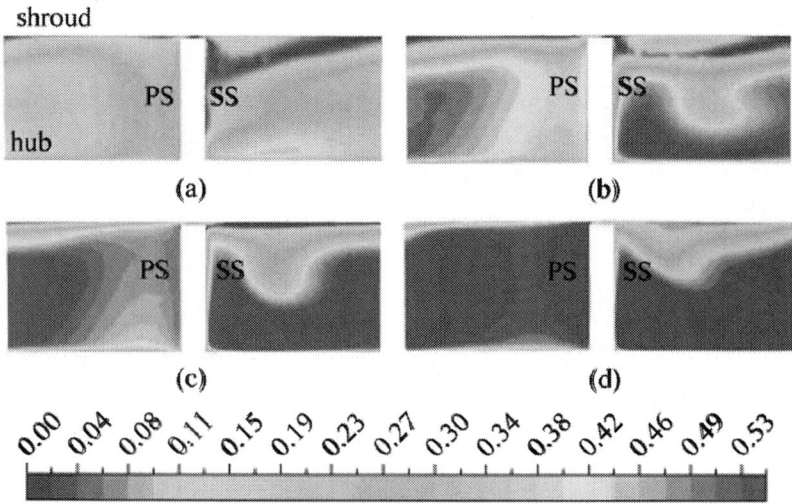

Figure 11: Contours of non-dimensional meridional velocity inside impeller at M = 0.97; R_3 = 1.05; (impeller with VD). (a) m = 0.423 kg/s; (b) m = 0.604 kg/s; (c) m = 0.805 kg/s; (d) m = 1.074 kg/s.

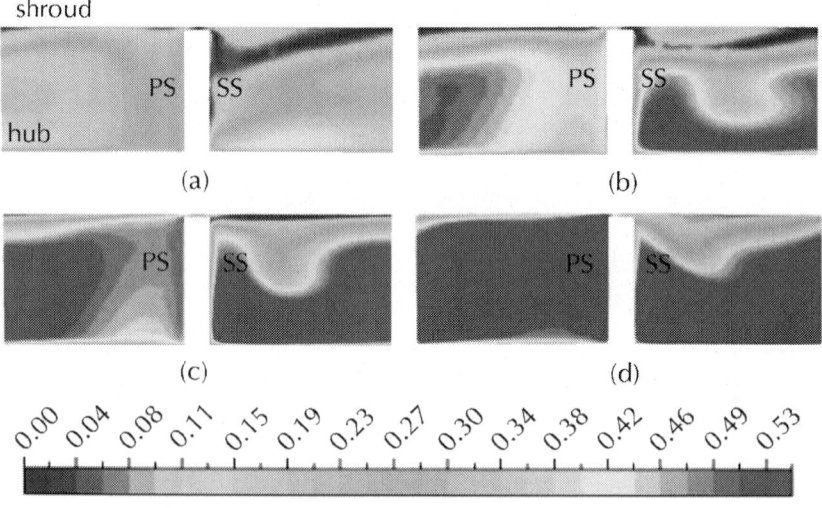

Figure 12: Contours of non-dimensional meridional velocity inside impeller at M = 0.97; R_3 = 1.15; (impeller with VD). (a) m = 0.423 kg/s; (b) m = 0.604 kg/s; (c) m = 0.805 kg/s; (d) m = 1.074 kg/s.

A look at the incidence angle plot (Figure 15) reveals that most of the span the incidence is negative at m = 0.805. Large positive incidence near the shroud is not observed at m = 0.805. Hayami et al. [12] found that a negative incidence of 2° to 3° gives better performance to low solidity vaned diffuser. Similarly Hohlweg et al. [13] also reported that the maximum diffuser performance is obtained with a negative incidence of 4.1°. Here it is found that small negative incidence is desirable in order to reduce the losses at the impeller exit. A definite value for the optimum incidence cannot be possible as the flow angle varies largely along the span.

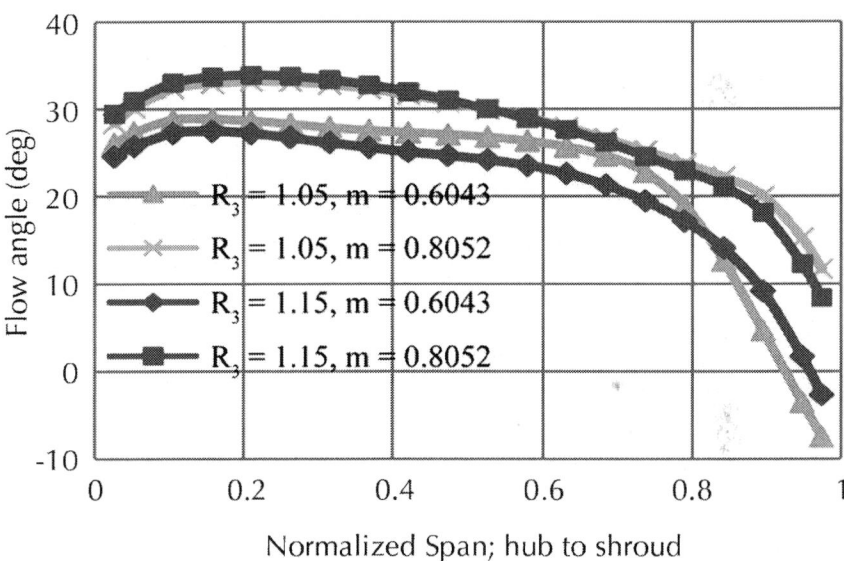

Figure 13: Circumferentially mass averaged flow angle in span-wise direction (0 = hub, 1 = shroud).

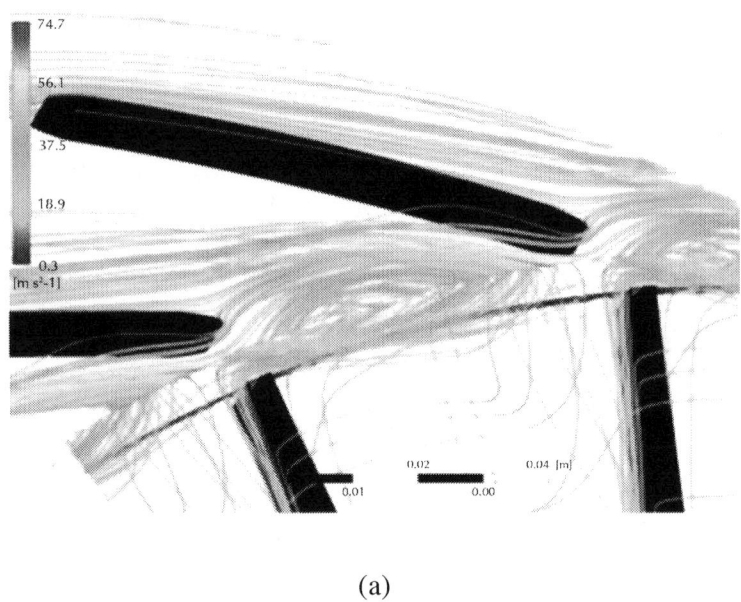

(a)

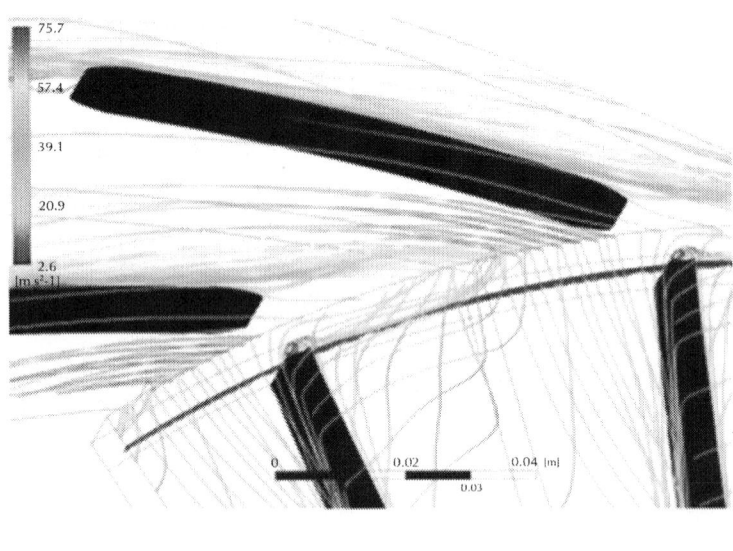

(b)

Figure 14: Streamlines show the flow behavior inside the diffuser for VD at R_3 = 1.05. (a) m = 0.604 kg/s; (b) m = 0.805 kg/s.

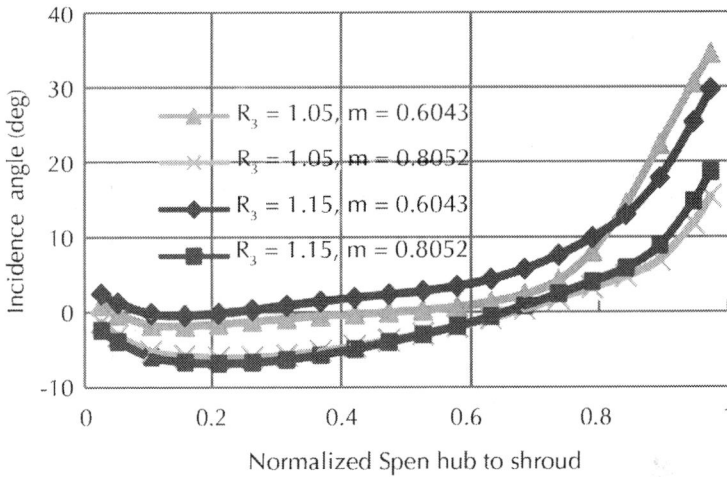

Figure 15: Circumferentially mass averaged incidence angle in span-wise direction (0 = hub, 1 = shroud).

Flow Behavior at Impeller Exit (LSVD and PVD)

The flow behavior inside the impeller with LSVD and PVD are dealt separately because the relative angular positions of impeller blade and diffuser vane are different from that of VD (ref. Figure 1).

Figure 16 shows the non-dimensional meridional velocity contours at $R_3 = 1.05$ and at $R_3 = 1.15$ respectively for LSVD. These are shown for a mass flow rate of m = 0.423 kg/s. There are notable difference regarding the strength and location of wake region between these two configurations. At $R_3 = 1.05$ wake region is located at the mid-passage close to the shroud. For $R_3 = 1.15$ the core of the reversed flow region is located at the shroud-suction surface corner. The streamline plot (Figure 17) give an insight into the flow behavior in these two configurations. When the diffuser vanes are closer to the impeller the reverse flow enhances the tip leakage flow near the impeller exit. This can be clearly observed from the streamlines pattern. Corresponding static entropy contours shows an increase in the entropy generation at this location (not shown here). When there is sufficient radial gap, the separated flow from the diffuser vane does not influence the impeller exit flow. Hence the losses at the impeller exit are also smaller.

The non-dimensional meridional velocity contours for the two PVD configurations are shown inFigure 18. The partial vaned diffuser does not affect the impeller exit flow adversely. On the contrary it stabilizes the flow near the shroud. Comparing the exit flow conditions of PVD with VLD reveals that the strength of the reverse flow region is reduced with the use of PVD.

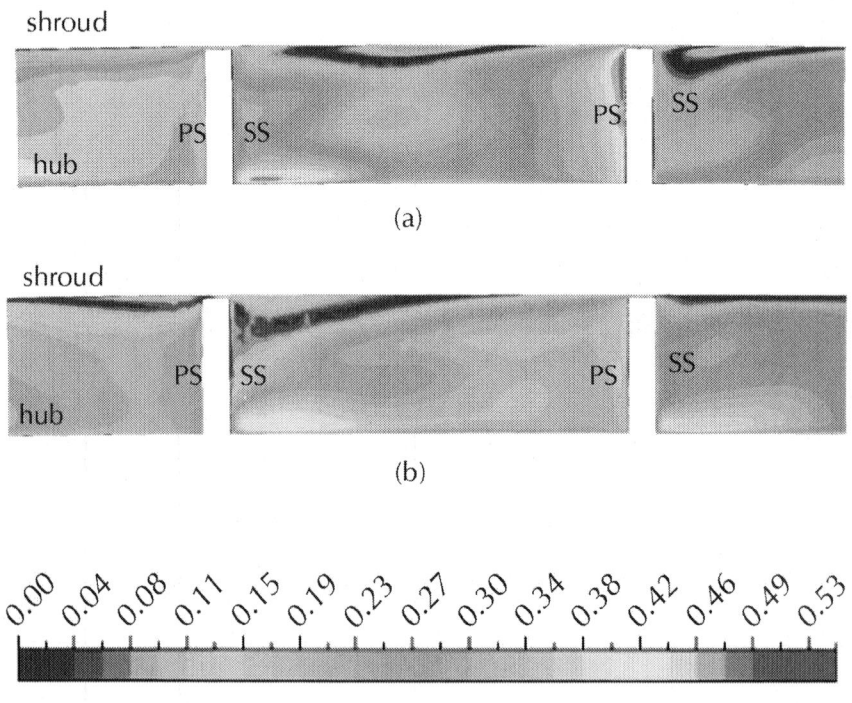

Figure 16: Contours of non-dimensional meridional velocity inside impeller (LSVD) at M = 0.97; m = 0.423 kg/s. (a) R_3 = 1.05; (b) R_3 = 1.15.

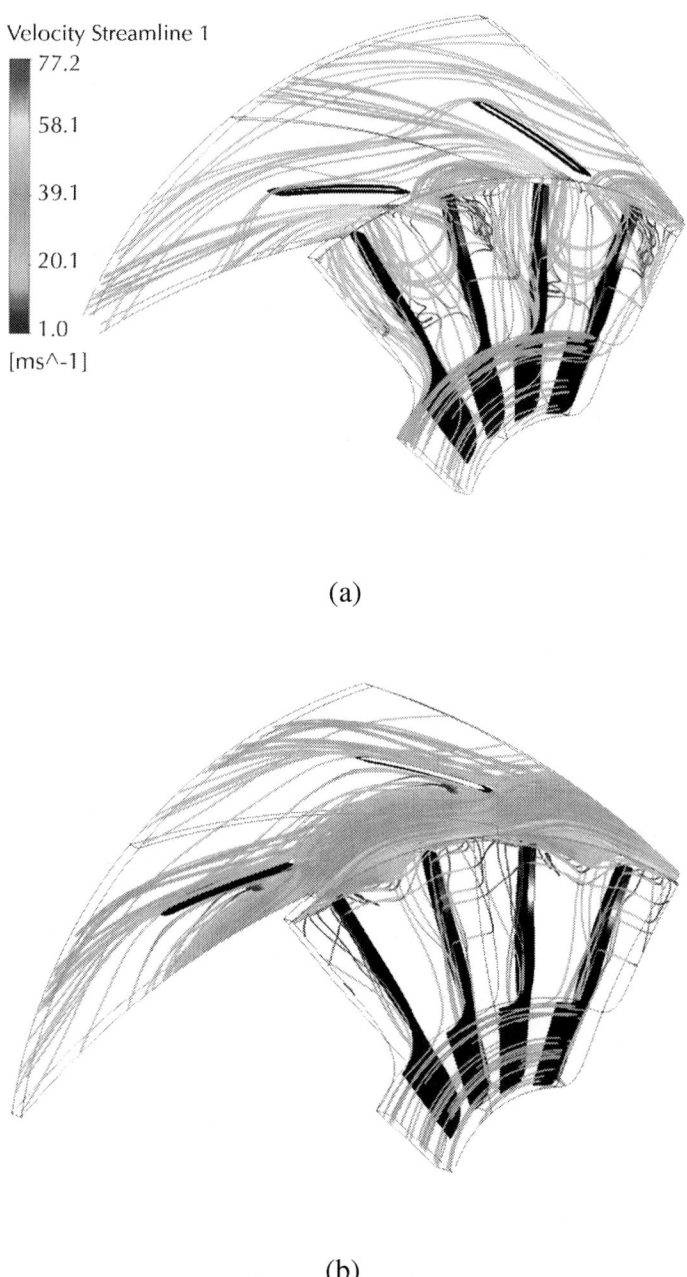

(a)

(b)

Figure 17: Streamline plots showing enhanced tip leakage flow for $R_3 = 1.05$. (a) $R_3 = 1.05$; (b) $R_3 = 1.15$.

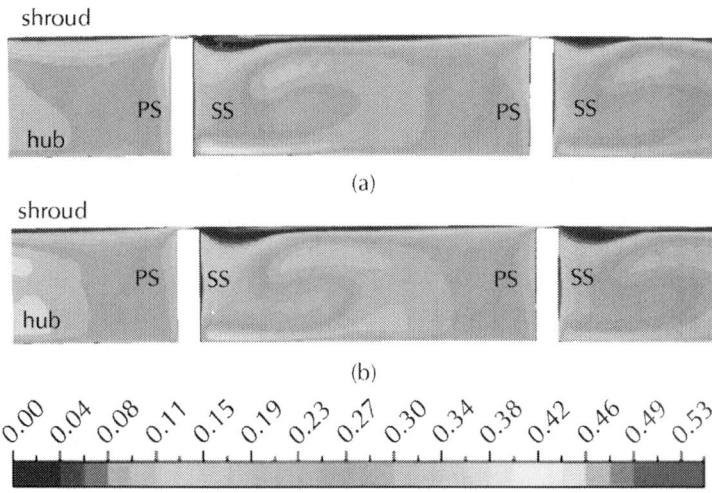

Figure 18: Contours of non-dimensional meridional velocity at M = 0.97, impeller with PVD; m = 0.4227 kg/s. (a) R_3 = 1.05; (b) R_3 = 1.15.

CONCLUSIONS

A computational study has been conducted to analyze the effects of diffuser vane on impeller exit flow. Three types of vane diffusers are used; a conventional vaned diffuser (VD), low solidity vaned diffuser (LSVD) and a partial vaned diffuser (PVD). The diffuser vane leading edge is kept at different radial locations to understand its influence on the impeller exit flow. Above design flow rate, the diffuser vane affects only the last twenty percent of the impeller flow. However below design flow rate, keeping vane closer to the impeller can cause an early stall within the impeller. Detailed flow analysis inside the impeller passage shows that the strength and location of the wake region at the exit of impeller blade is heavily depended upon the tip leakage flow and the pressure equalization flow.

Even at design flow rate the flow angle near the shroud is too small and hence it causes large positive incidence. This enhances the reverse flow near shroud. A small negative incidence at the diffuser vane is desirable in order to reduce the losses at the impeller exit. Partial vane diffusers do not affect the impeller exit flow adversely instead they stabilize the impeller separation region at lower flow rates.

REFERENCES

1. M. Inoue and N. A. Cumpsty, "Experimental Study of Centrifugal Impeller Discharge Flow in Vaneless and Vaned Diffusers," ASME Journal of Engineering Gas Turbines Power, Vol. 106, No. 2, 1984, pp. 455-467. doi:10.1115/1.3239588

2. J. Moore and J. G. Moore, "Calculation of Three-Dimensional, Viscous Flow and Wake Development in Centrifugal Impeller," Journal of Engineering for Gas Turbines and Power, Vol. 103, No. 2, 1981, 6 pp.

3. J. D. Denton, "The Use of a Distributed Body Force to Simulate Viscous Effects in 3D Flow Calculations," ASME 31st International Gas Turbine Conference and Exhibit, Duesseldorf, 8-12 June 1986, 8 pp.

4. W. N. Dawes, "A Simulation of the Unsteady Interaction of a Centrifugal Impeller with Its Vaned Diffuser: Flows Analysis," ASME Journal of Turbomachinery, Vol. 117, No. 2, 1995, pp. 213-222.

5. R. M. Chriss, M. D. Hathaway and J. R. Wood, "Experimental and Computational Results from the NASA Lewis Low-Speed Centrifugal Impeller at Design and Part-Flow Conditions," ASME Journal of Turbomachinery, Vol. 118, No. 1, 1996, pp. 55-65.

6. Ch. Hirsch, S. Kang and G. Pointel, "A Numerically Supported Investigation of the 3D Flow in Centrifugal Impellers, Part 1: Validation Base," ASME Paper No. 96-GT-151, ASME TURBO EXPO, Birmingham, 1996.

7. S. Kang and Ch. Hirsch, "Numerical Simulation and Theoretical Analysis of the 3D Viscous Flow in Centrifugal Impellers," TASK Quarterly, Vol. 5, No. 4, 2001, pp. 433-458.

8. M. Schleer and R. S. Abhari, "Clearance Effects on the Evolution of the Flow in the Vaneless Diffuser of a Centrifugal Compressor at Part Load Condition," ASME Journal of Turbomachinery, Vol. 130, 2008, pp. 1-9.

9. D. Eckardt, "Detailed Flow Investigations within a High Speed Centrifugal Compressor Impeller," ASME Journal of Fluids Engineering, Vol. 98, No. 3, 1976, pp. 390-402. doi:10.1115/1.3448334

10.   Y. K. P. Shum, C. S. Tan and N. A. Cumpsty, "Impeller-Diffuser Interaction in a Centrifugal Compressor," ASME Journal of Turbomachinery, Vol. 122, No. 4, 2000, pp. 777-786. doi:10.1115/1.1308570

11.   J. M. Issac, "Performance and Flow Field Measurements in Different Types of Diffusers of a Centrifugal Compressor," Ph.D. Thesis, Department of Mechanical Engineering, Indian Institute of Technology (IIT) Madras, Chennai, 2004.

12.   H. Hayami, Y. Senoo and K. Utsunomiya, "Application of a Low Solidity Cascade Diffuser to Transonic Centrifugal Compressor," ASME Journal of Turbomachinery, Vol. 112, No. 1, 1990, pp. 25-29. doi:10.1115/1.2927416

13.   W. C. Hohlweg, G. L. Direnzi and R. H. Aungier, "Comparison of Conventional and Low Solidity Vaned Diffusers," ASME Paper 93-GT-98, ASME TURBO EXPO, Cincinnati, 1993.

Chapter 3

Transient Hydraulic Performance and Numerical Simulation of a Centrifugal Pump with an Open Impeller during Shutting Down

Yuliang Zhang[1], Zuchao Zhu[1, 2], Yingzi Jin[2],
and Baoling Cui[2]

[1]The State Key Laboratory of Fluid Power Transmission and Control, Zhejiang University, Hangzhou, China
[2]The Province Key Laboratory of Fluid Transmission Technology, Zhejiang Sci-Tech University, Hangzhou, China

ABSTRACT

In this paper, the transient behavior of a low specific speed centrifugal pump with straight blades during shutting down is studied through the experimental test, theoretical calculation, and numerical simulation. The variations of the rotational speed, flow rate, and head with time are obtained in experiment. Based on the experimental results of the rotational speed and flow rate, the additional theoretical heads are quantitatively calculated and analyzed. The experimental results of the rotational speed and flow rate are worked as the boundary conditions to accurately accomplish the numerical simulation of the transient flow during shutting down. The experimental results show that the decrease history of the flow rate evidently lags behind that of the rotational speed, while the rotational speed slightly lags behind the head. Theoretical analysis shows that there exists a clear negative head impact phenomenon in the process of stopping. The transient behavior of the centrifugal pump with straight blades mainly comes from the rotation deceleration of impeller, and has nothing to do with the fluid deceleration. The numerical simulations show that a large area backflow can be seen when the rotational speed decreases to zero due to the flowing inertia. In conclusion, the numerical simulation of the flow field is in good agreement with the internal flow theory of centrifugal pumps.

INTRODUCTION

Centrifugal pumps are liable to malfunction when working at unstable states, such as startup and stopping. Meanwhile, the study on transient behavior is very limited by far. The main characteristics of these transient problems are as follows: very short time, intense variation, and unsteady flow. Startup and shut down of fluid machinery are unsteady transient flow problems caused by the movement of the wall. Some researchers have shown that transient effect is very obvious during startup [1-6], the distinction between stable and unstable states is evident. Presently, the research on transient behavior has basically concentrated on the startup process of centrifugal and mixed flow pumps, while the studies of the transient behavior of shut down are few. Wang [7] researched the shutdown of a mixed flow pumps. These results show the head, flow rate and energy of the piping system rapidly decline during shut down, and proposed an operation method for shut down. Tsukamoto [8] researched the shutdown of a centrifugal pump by experimental test and theoretical analysis, his results show that the initial pressure coefficient is greater than falls below the quasistable state value.

With the development of wider application field, it is more necessary to study the interior flow characteristics of centrifugal pumps to reveal their transient performance. As is well known, the instantaneous rotational speed, flow rate, and pressure quickly vary with time during shut down. If these conditions are worked as boudary conditions in simulation, the flow calculation can be accomplished. Therefore, in this work, a calculation mothod is adopted—numerical simulation based on experimental results. In this method, the rotational acceleration is taken into account and experimental results (rotational speed, flow rate etc.) are defined as time functions, which are then written to program by user defined function (UDF). The method can accomplish a precise numerical simulation of transient problem. For the time functions, the rotational speed and flow rate are indispensable.

HYDRAULIC TEST

Experiment Equipment

Test pump is a centrifugal pump with small flow rate, and belongs to super-low-specific-speed range, whose impeller is open and designed with straight blades. Design parameters are as below: flow rate is 2.5 m³/h, head is 120 m, rotational speed is 2900 rpm. Hydraulic tests are carried out by closed-type test rig with B grade precision. Experimental equipments are shown in Figure 1. Power source is the variable-frequency motor, the transient flow rate is measured by LWGY type turbine flowmeter, the transient rotational speed is measured by JC1A rotational speed sensor and JW-2A microcomputer torque meter. Acquisition and disposal of test datum are accomplished by PCI-6023E data acquisition card and LabVIEW virtual instrument platform. Additionally, a KF1851 electric capacity type differential pressure transmitter is also assembled, which may be used to measure the pressure at inlet and outlet of the pump.

Results and Discussion

During shut down, centrifugal pump abruptly loses the driving power, and impeller begins to decelerate from a stable rotational speed to zero in a very short time. In this process, each parameter would change severely. Due to inertia effects, the rotational speed of impeller does not stop immediately, namely there exists a slowing process. Flow would last for a time interval, so its flow rate does not also abruptly reach zero. In short, with the deceleration of the impeller, the flow rate rapidly declines. When the stable rotational speed is 2930 rpm and the valve opening is 1.44 times rated flow rate, the experimental result is shown in Figure 2.

Result shows that the rotational speed, head, and flow rate rapidly decline with the time during shut down. At the beginning of shut down, three parameters fast decline, and then the speed rate of decrease becomes slow. The attenuation curves present the characteristics of polynomial function.

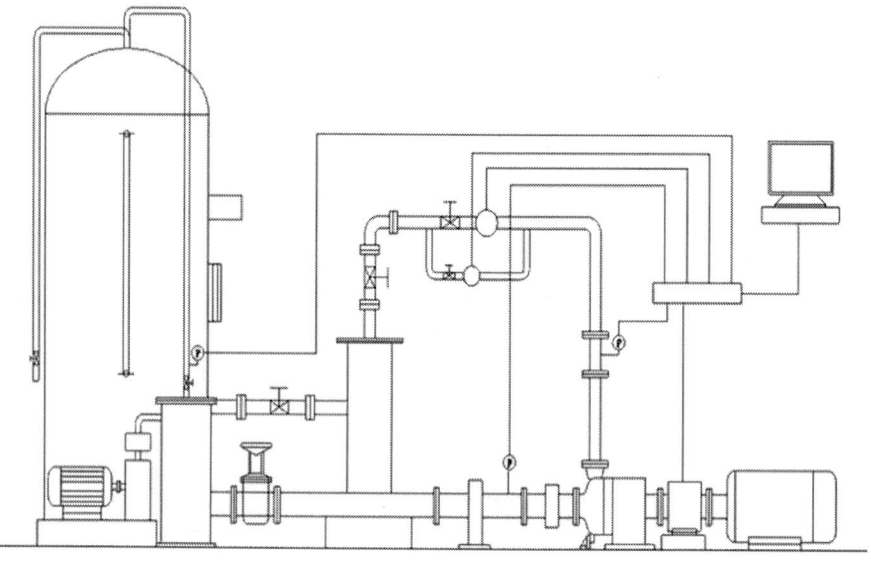

Figure 1: Test rig of external performance.

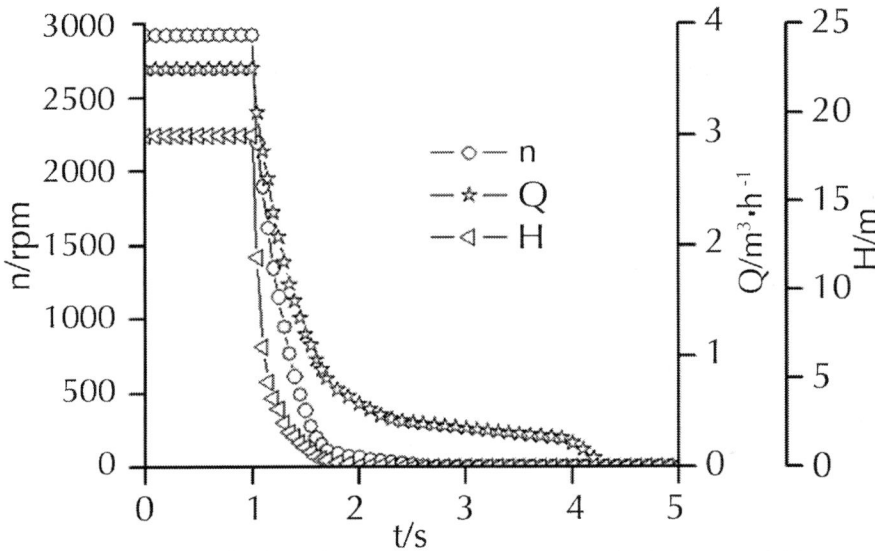

Figure 2: Variations of external performance during shutting down.

It takes about 0.6 s to accomplish 93% decelerating in the rotational speed. Similarly about 1 s to accomplish 84% decline in the flow rate and 99% decline in the head. For the flow rate and head, it takes about 0.6 s to accomplish 73% and 96%, respectively. The flow rate evidently lags behind the decline process of the rotational speed, while the rotational speed also lags behind that of the head. When the rotational speed declines to zero, the inertia effect would keep fluid flowing, so the flow rate lags behind the rotational speed. The stopping time in head, rotational speed and flow rate are 1.0 s, 2.1 s and 3.3 s, respectively. The following reasons may explain above results: the characteristic of driving power mainly determines decelerating process of the rotational speed. The higher rotational speed, the more surplus energy, so the time spent during shut down would be longer. Besides pump, the decline process of flow rate is closely relates to the piping system, more resistance, less time.

THEORETICAL CALCULATION OF TRANSIENT HEAD

Generalized Equation

According to the theorem of moment of momentum, the universal equation of centrifugal pump is as below [9].

$$H_d = \frac{u_2 v_{u2} - u_1 v_{u1}}{g} + \frac{\omega}{gQ_d} \times \iiint_{\Omega} \frac{\partial (v_u r)}{\partial t} \, \mathrm{d}W_i$$

(1)

The first item in Equation (1) is Euler equation at stable state. The second item is the additional head during transient operation, which would bring transient behavior. In any transient process, angular velocity and flow rate only relate to time, while not relate to coordinate. Therefore

$$\rho \iiint_{\Omega} \frac{\partial (v_u r)}{\partial t} \, \mathrm{d}W_i = \rho \frac{\mathrm{d}\omega}{\mathrm{d}t} \iiint_{\Omega} r^2 \, \mathrm{d}W_i$$

$$- \rho \frac{\mathrm{d}Q_d}{\mathrm{d}t} \iiint_{\Omega} \frac{r}{F \cdot \mathrm{tg}\beta} \, \mathrm{d}W_i$$

(2)

Where

$$\left. \begin{array}{l} \rho \dfrac{\mathrm{d}\omega}{\mathrm{d}t} \displaystyle\iiint_{\Omega} r^2 \, \mathrm{d}W_i = \Omega_J \cdot D^5 \cdot \dfrac{\mathrm{d}\omega}{\mathrm{d}t} \\[2em] \rho \dfrac{\mathrm{d}Q_d}{\mathrm{d}t} \displaystyle\iiint_{\Omega} \dfrac{r}{F \cdot \mathrm{tg}\beta} \, \mathrm{d}W_i = \Omega_M \cdot D^2 \cdot \dfrac{\mathrm{d}Q_d}{\mathrm{d}t} \end{array} \right\}$$

(3)

Where D is the nominal diameter of impeller. In centrifugal impeller, $D = D_2$. The rotational inertia coefficient and the flow inertia coefficient of fluid in impeller are respectively as below.

$$\Omega_J = \frac{\pi\rho}{32}\left(\bar{D}_2^4\bar{b}_2 - \bar{D}_1^4\bar{b}_1\right)$$

$$\Omega_M = \frac{\rho}{8}\left(\frac{\bar{D}_2^2}{\psi_2 tg\beta_2} - \frac{\bar{D}_1^2}{\psi_1 tg\beta_1}\right)$$

(4)

where the meaning of each parameter in (4) can be seen in Ref. [9]. The additional heads of centrifugal pump during transient operating periods are as below.

$$H_u = H_{u1} - H_{u2}$$

(5)

$$H_{u1} = \frac{\omega}{\rho g Q_d} \cdot \Omega_J \cdot D^5 \cdot \frac{d\omega}{dt}$$

$$H_{u2} = \frac{\omega}{\rho g Q_d} \cdot \Omega_M \cdot D^2 \cdot \frac{dQ_d}{dt}$$

(6)

where the former in (6) is the additional head brought by the varying rotational speed, the later is another additional head brought by the varying flow rate.

Transient Head

In this paper, main geometry parameters of centrifugal impeller are shown in Table 1.

According to the experimental results, geometry value in Table 1, and transient theory head Equation (6), the flow inertia coefficient of fluid in impeller can be written.

$$\Omega_M \approx 0$$

(7)

The result shows that transient behavior of centrifugal pump with straight blades mainly comes from the rotation deceleration of impeller and has little to do with the fluid deceleration. Figures 3 and 4 show the theoretical results.

It is seen that the additional theoretical head is of a clear negative impulse phenomena. The influence of the rotation acceleration of impeller (actually a negative value) on theory head is very evident. In this process, the influence of the fluid deceleration on theory head is negligible. Due to declining flow rate, the passing relative velocity in impeller channel is also low. Rotational speed declines faster and the fluid inertia generates a large area backflow region, which causes plenty of hydraulic loss and makes head decline quickly. The decline process of static theory head is nearly consistent with that of rotational speed. The decrease head brought by the rotation deceleration mainly comes from initial stage of impeller decelerating, which is likely to relate to too high negative acceleration.

Table 1: Main geometric paramter of impeller

Geometric Parameter	Symbol	Value
Number of blades	Z	8
Impeller inlet diameter	D_1	50 mm
Impeller outlet diameter	D_2	270mm
Impeller inlet width	b_1	17mm
Impeller outlet width	b_2	5mm
Blade angle at inlet	β_1	90°
Blade angle at outlet	β_2	90°
Emission coefficient at inlet	ψ_1	0.91
Emission coefficient at outlet	ψ_2	0.96
Nominal diameter	D	270mm

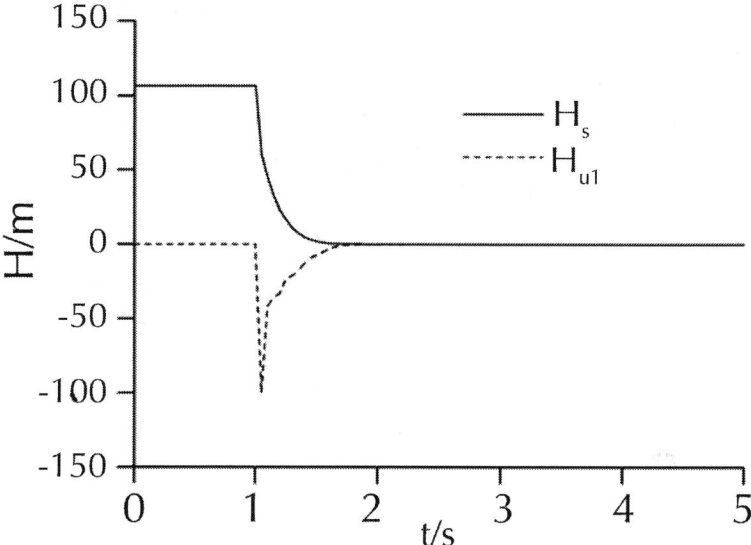

Figure 3: Transient heads during shut down.

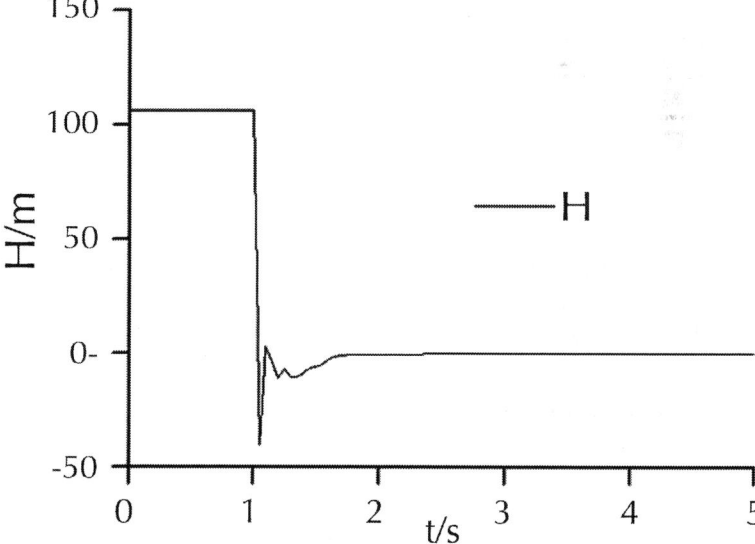

Figure 4: Total theoretical head during shut down.

NUMERICAL SIMULATION

Solving Control

To successfully solve the transient flow during shutting down, the flow domain is divided into two region: dynamic and stationary. The governing equation in the dynamic region are described in the arbitrary Lagrangian and Eulerian way (or the dynamic mesh conservation equations described in the FLUENT User's Guide), whereas the stationary region is governed by ordinary N-S equation. Different regions are separately discretized, so the non-conformal meshes between two regions are linked by a pair of mesh interfaces described in the FLUENT User's Guide. The sliding mesh technique is used to accomplish data exchange between rotor and stator.

For the flow in the dynamic region, the integral form of the dynamic mesh conservation equations for space, mass, and momentum in an arbitrary control volume V bounded by a closed surface S can be written as

$$\frac{d}{dt}\int_V Q dV + \int_S F \cdot dS = \int_S D \cdot dS + \int_V S_u dV \tag{8}$$

where Q is the conservation variable, F is the convective flux, D is the diffusion term, and S_u is the source term of the momentum. They are given in the following expressions:

$$Q = \begin{bmatrix} 1 \\ \rho \\ \rho u \end{bmatrix} \quad F = \begin{bmatrix} -u_b \\ \rho(u-u_b) \\ \rho u(u-u_b) \end{bmatrix}$$

$$D = \begin{bmatrix} 0 \\ 0 \\ \mu\nabla^2 u \end{bmatrix} \quad S_u = \begin{bmatrix} 0 \\ 0 \\ -\text{grad } p \end{bmatrix} \tag{9}$$

where ρ is the fluid density, u is the fluid velocity, u_b is the boundary velocity of the moving mesh, μ is the dynamic viscosity of the fluid, and p is the pressure of the fluid.

Unsteady flow is calculated by means of CFD software—FLUENT that based on a finite volume method, Yakhoth and Orzag put forward RNG $k - \varepsilon$ turbulence model in 1986, it is well suited to interior flow of pumps and well-disposed for flows of high strain rate and high curve degree. In calculation region, multi-block mesh technology and local mesh refinement technology are used to acquire flow detail, and the total number of grids is 760,000. As is well known, the grid number is not enough for study the some micro flow like in the boundary layer, but is enough for predict the external performance and the basic flow characteristics. Open impeller and pump are shown in Figure 5. Time discretization of transient term adopts one order implicit scheme, space discretization of convection term and diffusion term adopts two order upwind scheme and central difference scheme with second order accuracy, space discretization of source term adopts standard scheme (linearization). The coupling between pressure and speed is accomplished by SIMPLE algorithm in simulations. The convergence criterion is 0.0001. The time step is set to 0.0001 s.

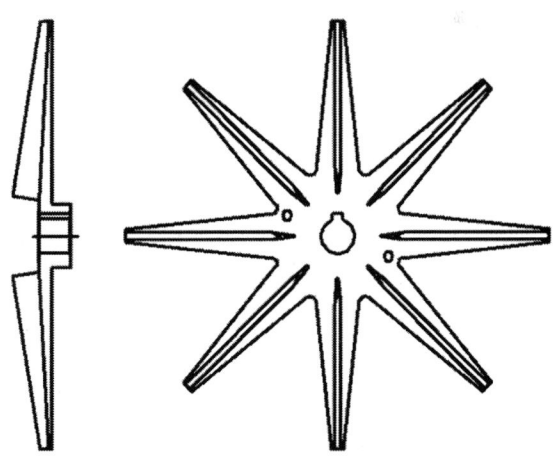

(a)

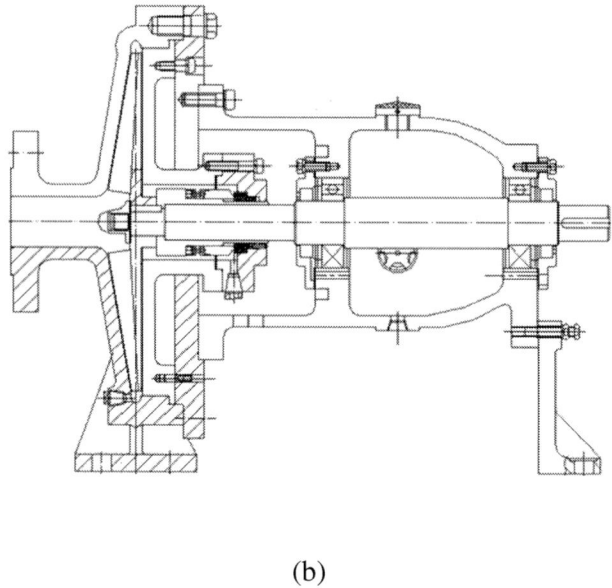

(b)

Figure 5: Pump model with open impeller. (a) Open impeller; (b) Pump sketch.

Boundary Condition

In this paper, time functions of the rotational speed and flow rate would be fit as the boundary condition of numerical simulation. To accomplish precise fitting of the test datum, the sectional fitting is applied.

- Rotational speed. Time histories of rotational speed are fit by exponential function.

$$n = 3.315 \times 10^5 \times e^{-\left(\frac{t+1.639}{1.209}\right)^2}$$

$$(10)$$

- Inlet condition. Time histories of flow rate are fit by polynomial function.

$$Q = -0.102t^5 + 1.53t^4 - 9.131t^3 + 27.13t^2 - 40.3t + 24.46$$

$$(11)$$

- Outlet condition. The assumption is that flow is fully developed, namely each parameter does not change along flow direction. Outflow condition is applied.

Wall condition: no slip boundary condition and standard wall function are adopted at wall.

Results and Analysis

Definition of static pressure coefficient is as below.

$$\psi = \frac{p}{2\rho u_2^2}$$

(12)

and

$$u_2 = \frac{\pi n D_2}{60}$$

(13)

therefore

$$\psi = \frac{1800\,p}{\rho \pi^2 n^2 D_2^2}$$

(14)

where p is static pressure, ρ is media density. Figure 6 is the distribution of static pressure coefficient on middle section of hub during shut down. Static pressure coefficients gradually decrease from working surface to back surface, while regularly increase from inlet to outlet. Action force and centrifugal force formed by pressure difference between working and back surface are consistent with the rotational direction of centrifugal pump, which completely accords with reality of centrifugal pump. Rotor-stator interaction makes the distribution of static pressure coefficient in channel vary obviously. Static pressure coefficient initially increases to a maximum at 0.5 s, before decreasing. At 1.2 s, static pressure coefficient begins to decrease and present regular character again. But volute tongue makes flow field present unsymmetrical structure. When time is 2 s, the rotational speed is very low, nearly reaches zero. Except for volute tongue region, distribution of static pressure coefficient is uniform.

Figure 7 is the distribution of turbulence intensity during shut down. Turbulence intensity is defined as a ratio of root mean square of fluctuation velocity to mean velocity. Its temporal and spatial distributions reflect conversion process of fluctuation velocity during shut down. During shut down, high turbulence intensity converts to low turbulence intensity, i.e., high Reynolds number (may reach to millions) converts low Reynolds number and till zero. Hence the turbulence mode would develop and change.

Figure 8 is the streamline on middle section of hub during shut down. At the beginning of startup, no backflow appears, then large area axial eddies appear and the rotational direction is reversed. When time is 2 s, rotational speed is very low and close to zero, but flow rate is still relatively high and backflow region still exists. Meanwhile, backflow region severely decreases and center of every axial eddy close to rotation center. Rotational speed has reached to zero after 3 s, then the flow around vanes leads to a backflow in channel. The Rossby dimensionless number could reflect the influence of vane curvature and rotation on backflow in the impeller. $Ros = w / 2\omega R$, this formula shows the ratio of inertial force (centrifugal force) to Coriolis force. Where curvature brings inertial force and rotation brings Coriolis force. For flow rates laging behind rotational speed, the Rossby number gradually increases during shut down. In Rossby number, the circular velocity is far lower than its relative velocity. Except above reason, valve type and annular volute may are another two important reasons in this paper.

(a)

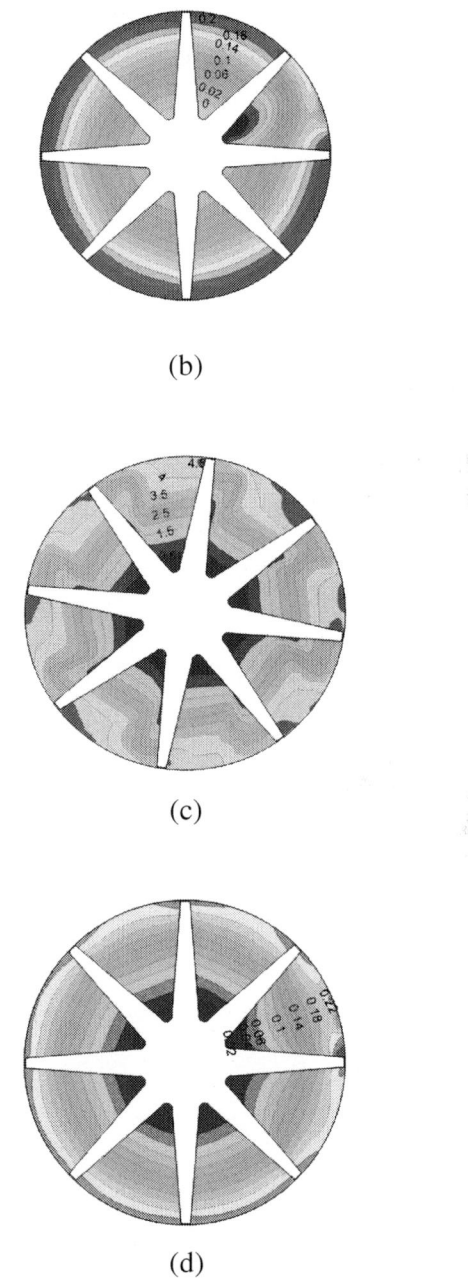

Figure 6: Distribution of static pressure coefficient during shutting down. (a) 0.1 s; (b) 0.5 s; (c) 1.2 s; (d) 2.0 s.

(a)

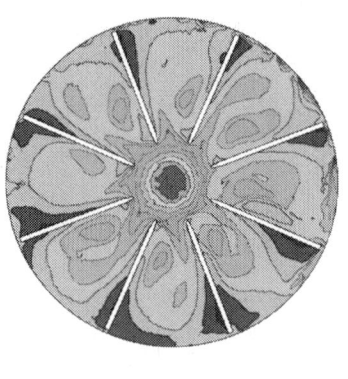

(b)

(c)

(d)

Figure 7: Variation of turbulence intensity during shut down. (a) 0.1 s; (b) 1.2 s; (c) 2.0 s; (d) 3.0 s.

(a)

(b)

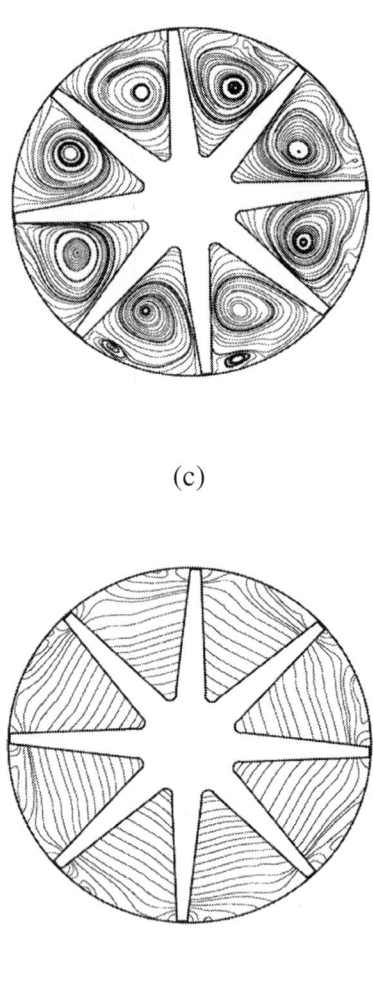

(c)

(d)

Figure 8: Variation of streamline during shutting down. (a) 0.1 s; (b) 1.2 s; (c)2.0 s; (d) 3.0 s.

CONCLUSIONS

Transient behavior of a low-specific-speed centrifugal pump with an open impeller and straight blades during shut down is researched. Experimental result approves that the decline of the flow rate evidently lags behind that of the rotational speed, which in turn lags behind that of the head, and three parameters decline rapidly initially before slowing. Theoretical analysis shows that transient theory head is of a clear negative-impulsion phenomena. Transient effect of centrifugal pump with straight blades mainly comes from the rotational acceleration of impeller, while the fluid acceleration is negligible. During shut down, the Rossby dimensionless number gradually increases. Flow inertia makes a large area backflow appear when the rotational speed approaches zero. Transformation trend of the flow field in numerical simulation reflects real flow characters of shut down, which are consistent with interior flow theory of centrifugal pumps.

ACKNOWLEDGEMENTS

This study is supported by the National Natural Science Foundation of China (No. 21076198, 51276172) and the National Basic Research Program ("973" Program, No. 2009CB724303).

REFERENCES

1.　H. Tsukamoto and H. Ohashi, "Transient Characteristics of a Centrifugal Pump during Starting Period," ASME Journal of Fluids Engineering, Vol. 104, No. 1, 1982, pp. 6-13. doi:10.1115/1.3240859

2.　P. J. Lefebvre and W. P. Barker, "Centrifugal Pump Performance during Transient Operation," ASME Journal of Fluids Engineering, Vol. 117, No. 1, 1995, pp. 123-128.doi:10.1115/1.2816801

3.　K. Farhadi, A. Bousbia-salah and F. D'Auria, "A Model for the Analysis of Pump Start-Up Transients in Tehran Research Reactor," Progress in Nuclear Energy, Vol. 49, No. 7, 2007, pp. 499-510. doi:10.1016/j.pnucene.2007.07.006

4.　P. Thanapandi and R. Prasad, "Centrifugal Pump Transient Characteristics and Analysis Using the Method of Characteristics," International Journal of Mechanical Sciences, Vol. 37, No. 1, 1995, pp. 77-89. doi:10.1016/0020-7403(95)93054-A

5.　A. Dazin, G. Caignaert and G. Bois, "Transient Behavior of Turbomachineries: Applications to Radial Flow Pump Startups,"

ASME Journal of Fluids Engineering, Vol. 129, No. 11, 2007, pp. 1436-1444. doi:10.1115/1.2776963

6.   S. Duplaa, O. Coutier-Delgosha, A. Dazin, et al., "Experimental Study of a Cavitating Centrifugal Pump during Fast Startups," ASME Journal of Fluids Engineering, Vol. 132, No. 2, 2010, Article ID: 021301.

7.   W. Chen, X. W. Ke, D. Z. Wu, et al., "Analysis on Transient Performance of Mixed Flow Pump During Stopping Period," Fluid Machinery, Vol. 34, No. 12, 2006, pp. 1-4.

8.   H. Tsukamoto, S. Matsunaga and H. Yoneda, "Transient Characteristics of a Centrifugal Pump during Stopping Period," ASME Journal of Fluids Engineering, Vol. 108, No. 4, 1986, pp. 392-399. doi:10.1115/1.3242594

9.   J. S. Chang, "Transients of Hydraulic Machine Installations," Higher Education Press, Beijing, 2005.

Chapter 4

Numerical Identification of Key Design Parameters Enhancing the Centrifugal Pump Performance: Impeller, Impeller-Volute, and Impeller-Diffuser

Massinissa Djerroud, Guyh Dituba Ngoma, and Walid Ghie

Department of Applied Sciences, University of Quebec in Abitibi-Témiscamingue, 445 Boulevard de l'Université, Rouyn-Noranda, QC, Canada J9X 5E4

ABSTRACT

This paper presents the numerical investigation of the effects that the pertinent design parameters, including the blade height, the blade number, the outlet blade angle, the blade width, and the impeller diameter, have on the steady state liquid flow in a three-dimensional centrifugal pump. Three cases were considered for this study: impeller, combined impeller and volute, and combined impeller and diffuser. The continuity and Navier-Stokes equations with the k- turbulence model and the standard wall functions were used by means of ANSYS-CFX code. The results achieved reveal that the selected key design parameters have an impact on the centrifugal pump performance describing the pump head, the brake horsepower, and the overall efficiency. To valid the developed approach, the results of numerical simulation were compared with the experimental results considering the case of combined impeller and diffuser.

INTRODUCTION

At the present time, single and multistage centrifugal pumps are widely used in industrial and mining enterprises. One of the most important components of a centrifugal pump [1] is the impeller. The performance characteristics related to the pump comprising the head, the brake horsepower, and the overall efficiency rely a great deal on the impeller. To achieve better performance for a centrifugal pump, design parameters such as the number of blades for the impeller and the diffuser, the impeller blade angle, the blade height for the impeller and the diffuser, the impeller blade width, the impeller diameter, and the volute radius must be accurately determined, due to the complex liquid flow through a centrifugal pump. This liquid flow is three-dimensional and turbulent. It is therefore important to be aware of the liquid flow's behavior when traveling through an impeller. This can be done by accounting for the volute and/or the diffuser in the planning, design, and optimization phases at conditions of design and off-design. Many experimental and numerical studies have been carried out on the liquid flow through a centrifugal pump [2–21], where the effects of the number of impeller blades on the pump's performance were examined experimentally in [11, 12]. The effects of the impeller outlet blade angle on the pump's performance were also investigated numerically [13, 14], using a CFD code and experimentally in [15]. In [16] the dynamic effects due to the impeller-volute interaction within a centrifugal pump were numerically investigated, whereas the effects of the volute on velocity and pressure fields were examined in [17, 18]. Additional experimental investigation carried out [19] consisted of measuring unsteady velocity, the pressure and flow angle at the centrifugal pump's impeller outlet, with and without volute casing. The liquid flow and head distribution within a centrifugal pump's volute were compared with the impeller's characteristics, without the volute casing. Moreover, two centrifugal pump impellers with different outlet diameters for the same volute were examined both experimentally and numerically [20], to evaluate the influence the radial gap between the impeller exit and the volute tongue had on the unsteady radial forces acting upon the impeller of a centrifugal pump with volute casing. Additionally, the effects of flow behavior in a centrifugal pump, whose diffuser was subjected to different radial gaps, were investigated numerically in [21] using a CFD code. The analysis of previous works clearly demonstrated that research results obtained are specific to the centrifugal pump design parameter values and thus cannot be generalized. In this work therefore a numerical study was performed using a finite volume method according to the CFX code [22] to gain further insight into the characteristics of the three-dimensional turbulent liquid flow through an impeller, a combined impeller and volute, and a combined impeller and diffuser, while also considering various flow conditions and pump design parameters: blade heights of 12 mm, 18 mm, and 24 mm; blade numbers of 5, 7, and 9 for the impeller and 5, 8, and 12 for the diffuser; outlet blade angles of 9°, 28°, and 60°, blade widths of 4 mm, 10 mm, and 15 mm; impeller outer diameters of 285 mm and 320 mm. The reference dimensions selected for the impeller and diffuser were based on the existing impeller and diffuser [23]. Upon applying the continuity and Navier-Stokes equations, the liquid flow velocity and the liquid pressure distributions in an impeller, a combined impeller and volute, and a combined impeller and diffuser were determined, while accounting for boundary conditions. Since the rotating speed of the centrifugal pump under consideration was constant a valve installed on the pump's discharge side was used to regulate the volume flow rate. We accounted for suction pressure variation as a function of the valve volume flow rate in the numerical simulations being run. The pump head, brake horsepower, and efficiency were represented as a function of the volume flow rate, where the objective was to identify the values of selected key design parameter that might improve pump performance with respect to their value ranges.

MATHEMATICAL FORMULATION

Figure 1 shows a centrifugal pump consisting of three components, including an impeller, a diffuser, and a volute [24]. The models selected for the liquid flow in an impeller, a combined impeller and volute, and a combined impeller and diffuser are depicted in Figure 2, placing greater emphasis on the fluid domain.

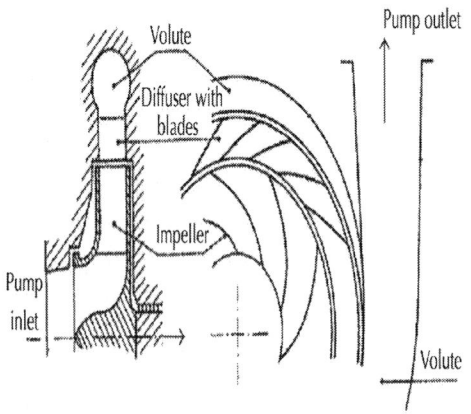

Figure 1: Centrifugal pump.

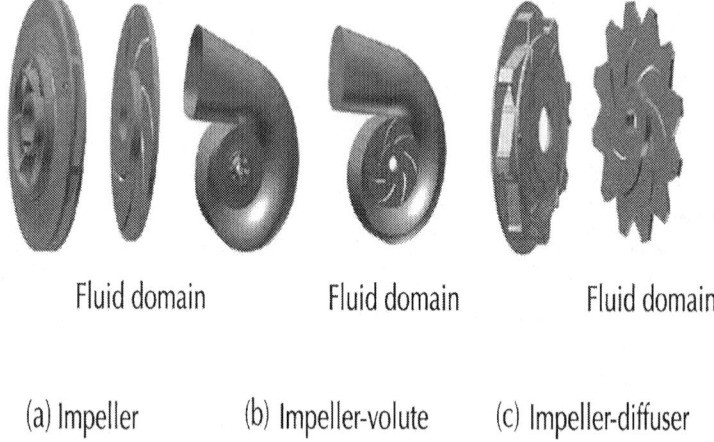

Fluid domain   Fluid domain   Fluid domain

(a) Impeller   (b) Impeller-volute   (c) Impeller-diffuser

Figure 2: Models of centrifugal pump components.

In the governing equations for liquid flow in the centrifugal pump components, the following assumptions were made: (i) a steady state, three-dimensional, and turbulence flow using the k- model; (ii) it was an incompressible liquid; (iii) it was a Newtonian liquid; (iv) the liquid's thermophysical properties were constant with temperature.

To account for these assumptions, the theoretical analysis of the liquid flow in an impeller, a combined impeller and volute, and a combined impeller and diffuser was based on the continuity and Navier-Stokes [22] equations. For the three-dimensional liquid flow through the components of a centrifugal pump as shown in Figure 2, the continuity equations are expressed by

$$\nabla \cdot \vec{U} = 0, \tag{1}$$

and the Navier-Stokes equations are given by

$$\rho \nabla \cdot \left(\vec{U} \otimes \vec{U} \right) = -\nabla p + \mu_{\text{eff}} \nabla \cdot \left(\nabla \vec{U} + \left(\nabla \vec{U} \right)^T \right) + B, \tag{2}$$

where $\vec{U} = \vec{U}$ (u(x, y, z), v(x, y, z), w(x, y, z)) is the liquid flow velocity vector, p is the pressure, ρ is the density, μeff is the effective viscosity accounting for turbulence, $\otimes$ is a tensor product, and B is the source term. More particularly, for flows in an impeller rotating at a constant speed ω, the source term can be written as follows:

$$B = -\rho \left(2 \vec{\omega} x \vec{U} + \vec{\omega} x (\vec{\omega} x \vec{r}) \right), \tag{3}$$

where $\vec{r}$ is the location vector.

In addition, μeff is defined as

$$\mu_{\text{eff}} = \mu + \mu_t,$$

(4)

where μ is the dynamic viscosity and μ_t is the turbulence viscosity.

Since the k-ε turbulence model is used in this work because convergence is better than with other turbulence models, μ_t is linked to turbulence kinetic energy k, equation (6), and dissipation ε, equation (7), via the relationship

$$\mu_t = C_\mu \rho k^2 \varepsilon^{-1},$$

(5)

where $C\mu$ is a constant.

The values for k and ε come directly from the differential transport equations for turbulence kinetic energy and turbulence dissipation rates:

$$\nabla \cdot \left(\rho \vec{U} k \right) = \nabla \cdot \left[\left(\mu + \frac{\mu_t}{\sigma_k} \right) \nabla k \right] + p_k - \rho \varepsilon,$$

(6)

$$\nabla \cdot \left(\rho \vec{U} \varepsilon \right) = \nabla \cdot \left[\left(\mu + \frac{\mu_t}{\sigma_\varepsilon} \right) \nabla \varepsilon \right] + \frac{\varepsilon}{k} (C_{\varepsilon 1} p_k - C_{\varepsilon 2} \rho \varepsilon),$$

(7)

where $C_{\varepsilon 1}$, $C_{\varepsilon 2}$, and σ_ε are constants. p_k is the turbulence production due to viscous and buoyancy forces, which is modeled using:

$$p_k = \mu_t \nabla \vec{U} \cdot \left(\nabla \vec{U} + \nabla \vec{U}^T \right) - \frac{2}{3} \nabla$$

$$\cdot U \left(3\mu_t \nabla \cdot \vec{U} + \rho k \right) + p_{kb},$$

$$p_{kb} = -\frac{\mu_t}{\rho \sigma_\rho} g \cdot \nabla \rho,$$

(8)

where p_{kb} can be neglected for the k-ε turbulence model.

Moreover, for the modeling of flow near the wall, the logarithmic wall function is used to model the viscous sublayer [22].

Impeller

Three velocity types are involved when considering the flow through a centrifugal pump impeller: the tangential velocity $U=r\omega$, the relative velocity W, and the absolute velocity V. The last is expressed in vector format as follows:

$$\vec{V} = \vec{U} + \vec{W}.$$

(9)

Figure 3 shows the velocity triangles at the impeller inlet and outlet at the design conditions where the liquid enters and leaves the impeller at the blade angles β_{b1} and β_{b2}, respectively. The components of $\overline{V}$ and $\overline{W}$ in the direction of $\overline{U}$ are V_u (swirl velocity) and W_u, respectively, while those normal to are $\overline{U}$ V_r and W_r.

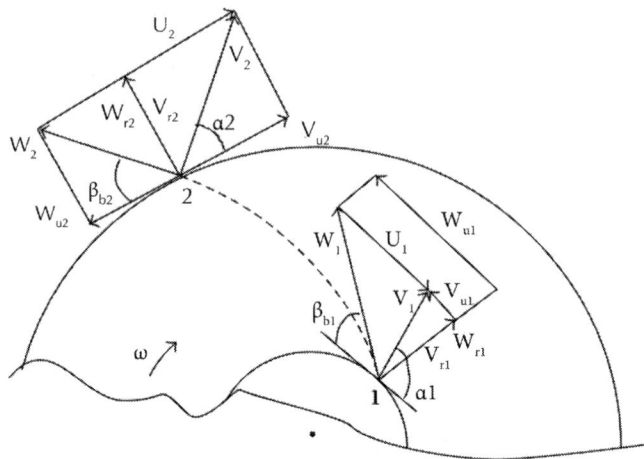

Figure 3: Velocity triangles.

Moreover, according to the Euler equation [1], the energy transfer per unit mass of flow for a centrifugal pump can be formulated as

$$gH_i = U_2 V_{u2} - U_1 V_{u1},$$

(10)

where H_i is the ideal pump total head.

Neglecting the swirl velocity at the impeller inlet, (10) can be expressed as follows:

$$gH_i = U_2 V_{u2}.$$

(11)

When accounting for the hydraulic efficiency, η_h, the actual pump head rise is given by

$$H = \eta_h H_i.$$

(12)

Also, the hydraulic efficiency can be calculated using the following empirical formula [1]:

$$\eta_h = 1 - \frac{0.8}{(15859.03Q)^{0.25}},$$

(13)

where Q is the volume flow rate in m³/s. It is given by $Q=V_rA$ with A as the flow passage area normal to the meridional direction.

Since in reality the flow through a centrifugal pump is turbulent and three-dimensional, the actual relative flow direction at the impeller exit is different from that of the blade angle. As depicted in Figure 4, the flow angle β_{f2} is always less than the blade angle β_{b2}. This can lead to secondary flows in the flow passage, from the pump inlet to discharge [1].

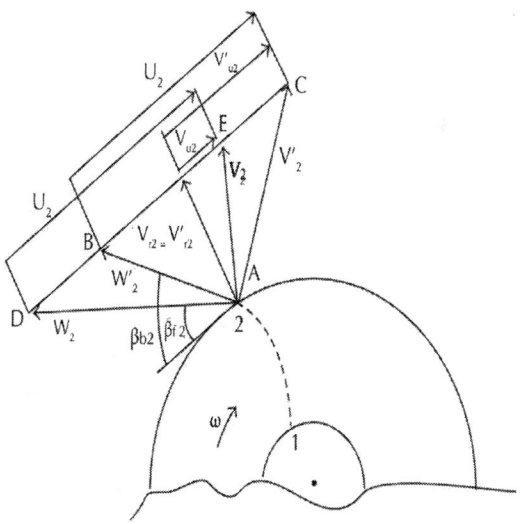

Triangle ABC : velocity triangle without slip

Triangle ADE : velocity triangle with slip

Figure 4: Flow angle and blade angle.

As such, the slip factor μ_s is used to take into account the difference between β_{b2} and β_{f2}, which is formulated as

$$\mu_s = \frac{V_{u2}}{V'_{u2}},$$

(14)

where V_{u2} is the actual swirl flow velocity at the impeller exit and V'_{u2} is the ideal swirl flow velocity at the impeller exit.

In addition, the slip velocity is given by

$$\Delta V_s = V'_{u2} - V_{u2} = W_{u2} - W'_{u2}.$$

(15)

Taking into account the slip factor, (12) can be expressed as

$$H = \eta_h \mu_s \left(\frac{U_2}{g} \right) \left(U_2 - \frac{Q}{A_2 \tan \beta_2} \right).$$

(16)

Moreover, to account for the leakage flow from the impeller, the volumetric efficiency is defined by

$$\eta_v = \frac{Q + Q_L}{Q_L},$$

(17)

where Q_L is the leakage flow from the impeller exit back to the inlet through the clearance.

In addition, the pump's mechanical efficiency is formulated as follows:

$$\eta_m = \frac{P_{imp}}{P_s},$$

(18)

where P_s is the brake horsepower and $Pimp$ the power delivered by the impeller to the fluid.

P_s is globally expressed by

$$P_s = P_h + P_f + P_L + P_m + P_{df} = C\omega,$$

(19)

where C is the pump shaft torque, P_h is the centrifugal pump horsepower. It is expressed as

$$P_h = \rho QgH.$$

(20)

P_f is the loss power due to the friction, which is given by

$$P_f = \rho Qg(H_i - H).$$

(21)

P_L is the loss power due to leakage, which is defined as:

$$P_L = \rho Q_L gH_i.$$

(22)

Pm is the friction loss power in bearings and seals, and P_{df} is the disk friction power due to impeller shrouds.

$Pimp$ in (18) can be formulated as follows:

$$P_{\text{imp}} = P_s - P_m - P_{df}.$$

(23)

Furthermore, (23) can be rewritten as

$$P_{\text{imp}} = \rho(Q + Q_L)gH_i.$$

(24)

Accounting for (23), (18) can be expressed as

$$\eta_m = \frac{P_s - P_m - P_{df}}{P_s}.$$

(25)

Thus, the overall efficiency of a centrifugal pump can be formulated as

$$\eta = \frac{P_h}{P_s}.$$

(26)

Finally, the overall efficiency can also be formulated in terms of the other efficiencies as

$$\eta = \eta_h \eta_v \eta_m.$$

(27)

Volute Parameters

Figure 5 shows the parameters of a volute without diffuser defined by the radius of volute basic circle r_3, the radius of volute cut water circle r_v, the volute angle α_v, the volute cross-sectional area A_θ, which depends on the angle , and the volute outlet cross-sectional area A_t [1].

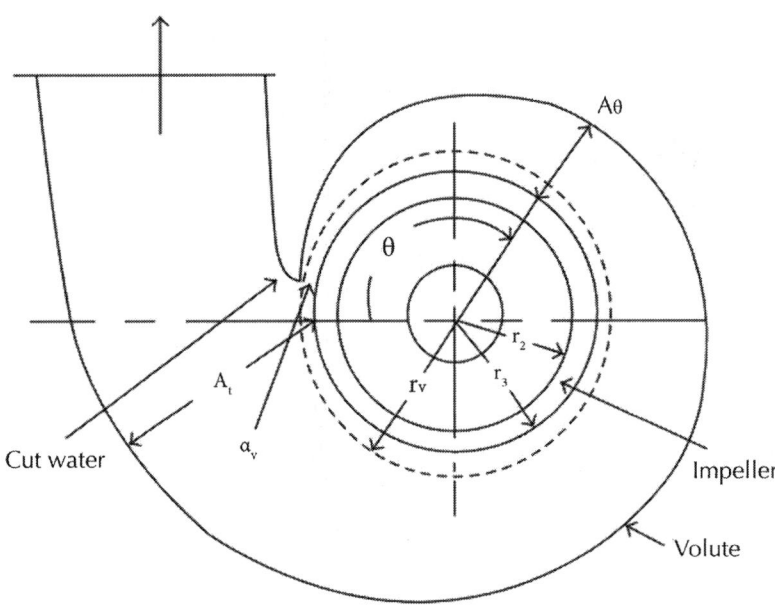

Figure 5: Impeller-volute without diffuser.

The average flow velocity at the volute outlet is given by

$$V_3 = K_3\sqrt{2gH},$$

(28)

where the volute velocity constant K_3 is an empirical parameter correlated with the specific speed, as shown in Figure 6 along with other volute parameters such as the volute angle α_v and the volute basic circle diameter D_3 [1].

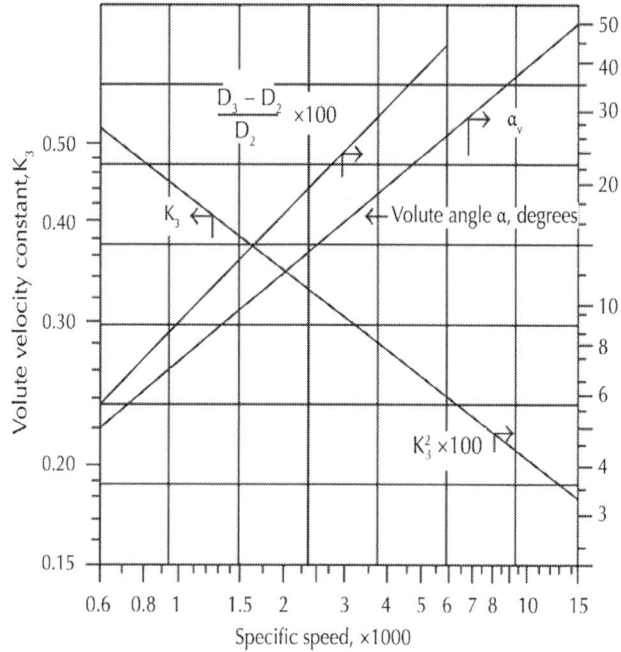

Figure 6: Volute velocity constant, volute angle, and diameter of volute basic circle versus specific speed.

In addition, the volute cross-sectional area A_θ can be formulated as

$$A_\theta = \frac{Q\theta}{2\pi CL} r_c,$$

(29)

where r_c is the centroid radius of the volute cross-sectional area, L is the angular momentum of flow at the impeller outlet, it is expressed as $L = r_2 V_{u2}$, and $C \cong 0.95$ to account for friction loss.

Diffuser Velocity and Pressure Difference

Figure 7 shows the velocities at the inlet and the outlet of a vaned diffuser immediately downstream from the impeller.

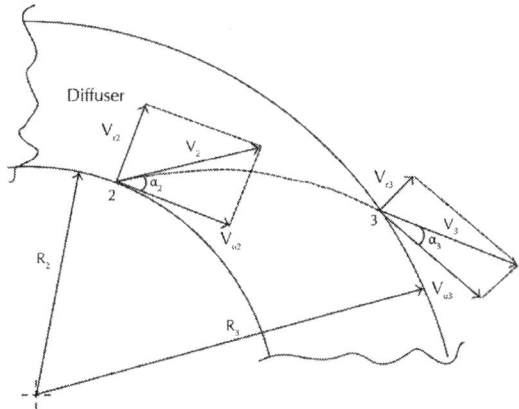

Figure 7: Velocities at the diffuser inlet and outlet.

The outlet velocity V_3 can be determined using [24]

$$V_3 = \sqrt{V_{u3}^2 + V_{m3}^2},$$

(30)

Where

$$V_{u3} = \frac{r_2 V_{u2}}{r_3}, \qquad V_{r3} = \frac{Q}{A_3},$$

(31)

where A_3 is the flow passage area normal to the meridional direction at the diffuser outlet.

Finally, the pressure difference between the diffuser outlet and inlet is given by

$$p_3 - p_2 = \frac{\rho}{2}\left(V_2^2 - V_3^2\right).$$

(32)

To solve (1) and (2) numerically while accounting for the boundary conditions and the turbulence model k- , the computational fluid dynamics ANSYS-CFX code, based on the finite volume method, was used to obtain the liquid flow velocity and the pressure distributions. In the cases examined involving the impeller, combined impeller and volute, and combined impeller and diffuser, the boundary conditions were formulated as follows: the static pressure provided was given at the inlet, while the flow rate provided was specified at the outlet. The frozen rotor condition was used for both the impeller-volute and the impeller-diffuser interfaces. A no-slip condition was set for the flow at the wall boundaries. Figure 8 shows the inlet, outlet, and interface domains for the selected centrifugal pump components.

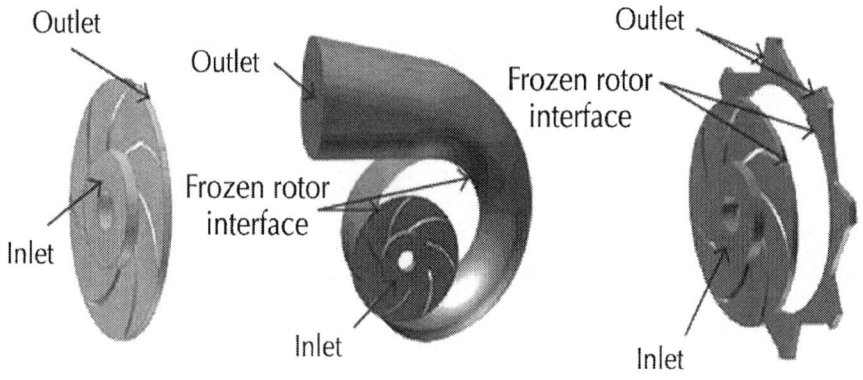

Figure 8: Domains of inlet, outlet, and interface.

Accounting for the fact that the pump rotating speed was constant, the volume flow rate was controlled by a regulator valve, which had an influence on the pressure at the pump inlet as indicated in Figure 9 [23]. This was accounted for in the numerical simulations performed.

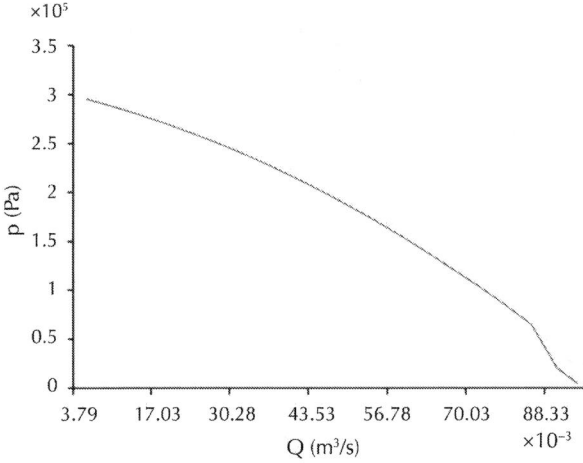

Figure 9: Pressure at the pump inlet versus valve volume flow rate regulation.

Furthermore, the ANSYS-CFX code comprises by geometry (DesignModeler), CFX-pre, CFX-solver, and CFX-post modules. According to the applied ANSYS-CFX code, Figure 10 depicts the steps specifically used to obtain the numerical simulation results from the geometry models to the numerical models for the impeller, the combined impeller and volute, and the combined impeller and diffuser.

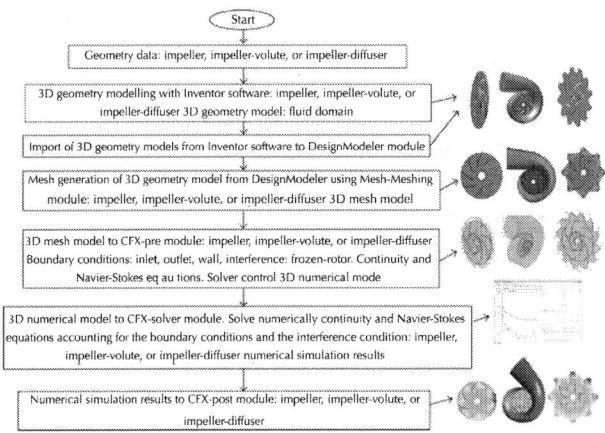

Figure 10: Steps from 3D geometry model to 3D numerical model and to numerical simulation results.

RESULTS AND DISCUSSION

Water was used as the working liquid for all simulations run and for use in this study considered to have the following reference values: temperature of 25°C for water, density of ρ = 997 kg/m³, and dynamic viscosity of μ= 8.899 × 10^{-4} Pa s. The main data for the reference impeller, volute, and diffuser are given in Tables 1, 2, and3.

Table 1: Main data of the reference impeller [23]

Inlet diameter (mm)	145
Outlet diameter (mm)	320
Inlet blade angle (degree)	11.69
Outlet blade angle (degree)	28
Inlet blade width (mm)	12
Blade thickness (mm)	4
Number of blades	7
Rotating speed (rpm)	1800

Table 2: Main data of the reference volute [1]

Volute angle (degree)	Volute radius (mm)	Volute angle (degree)	Volute radius (mm)
0	165	225	278.96
45	183.79	270	302.76
90	207.58	315	326.55
135	231.38	360	350.35
180	255.17		

Table 3: Main data of the diffuser [23]

Inlet diameter (mm)	321.536
Outlet diameter (mm)	441.77
Blade width (mm)	12

Blade thickness (mm)	3.401
Inlet blade angle (degree)	11.07
Outlet blade angle (degree)	39.42
Number of blades	8

Case Studies

Six key design parameters of a centrifugal pump were selected for an examination of their effects mainly on the pump performance: impeller blade height without volute, impeller blade width without volute, impeller blade angle without volute, impeller blade number with volute, impeller diameter with volute, and diffuser blade number with impeller. For the highest accuracy of numerical simulation results, mesh-independent solution tests were conducted in each case study by varying the number of mesh elements. Table 4 indicates the required number of mesh elements to achieve mesh-independent results.

Table 4: Number of mesh elements

Case study	Mesh element number
Impeller outlet blade height (2 in mm)	
12	43380
18	48776
24	74210
Impeller blade width (e in mm)	
4	43380
10	75841
15	119945
Impeller outlet blade angle (2 in degree)	
9	38508
28	43380
60	42288
Impeller blade number when accounting for volute	
5	28287

7	47979
9	132852
Impeller diameter when accounting for volute (2 in mm)	
285	47377
320	47979
Diffuser blade number	
5	39126
8	43380
12	45117

Effect of Impeller Outlet Blade Height

To analyze the outlet blade height's effect on the pump head, the pump brake horsepower, and the overall pump efficiency, the values 0.012 m, 0.018 m, and 0.024 m were selected for outlet blade height, while keeping the other parameters constant. Figure 11 shows the pump head as a function of the volume flow rate with the outlet blade height as a parameter. There, it can be clearly observed that the pump head decreases with increasing volume flow rate due to decreasing liquid pressure. In addition, the pump head increases with increasing outlet blade height. This can be explained by the fact that, when the volume flow rate is kept constant, the increased outlet blade height leads to the decreasing meridional velocity, which increases the pump head since the outlet tangential velocity and the outlet blade angle remain constant. In other words, the liquid pressure drop in the impeller decreases as a function of the increase in outlet blade height.

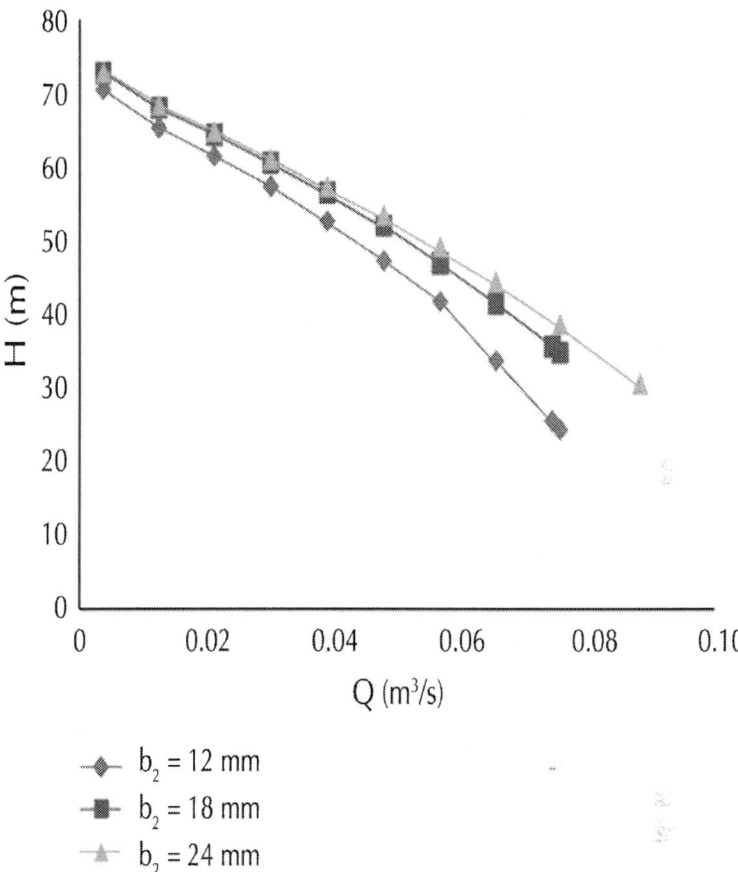

Figure 11: Pump head versus volume flow rate (parameter: blade height).

The curves expressing the pump brake horsepower as a function of the volume flow rate are shown in Figure12, illustrating that the brake horsepower increases with increasing volume flow rate. This can be explained by the additional decrease in liquid pressure relative to the volume flow rate. Also, the brake horsepower increases relative to the impeller blade height due to the requested increase in pump shaft torque relative to the increased blade height.

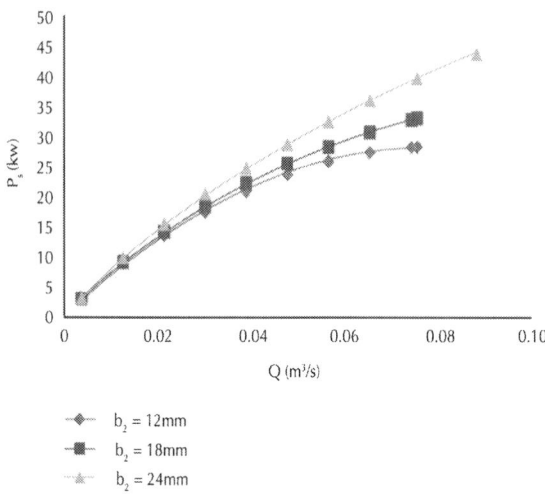

Figure 12: Brake horsepower versus volume flow rate (parameter: blade height).

As depicted in Figure 13, the curves representing overall pump efficiency as a function of volume flow rate illustrate that the overall efficiency for $b_2 = 12$ mm decreases rapidly to the right of the best efficiency point (BEP). The overall efficiency for $b_2 = 18$ is highest when the volume flow rate reaches 0.08 m³/s.

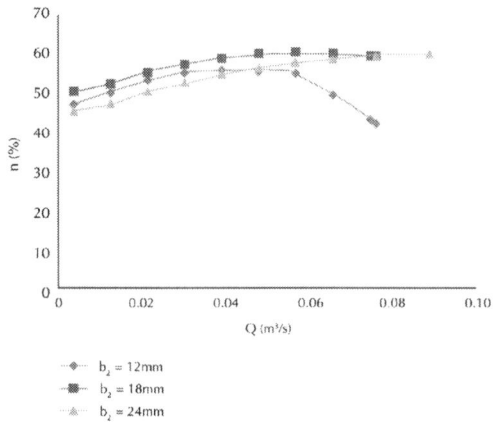

Figure 13: Overall efficiency versus volume flow rate (parameter: blade height).

Figures 14, 15, and 16 show the corresponding contours for static pressure, liquid flow velocity vectors, and streamlined liquid flow velocities for Q = 0.065 m³/s. From these figures it can be observed that the static pressure is higher at the impeller outlet than at the impeller inlet. This is due to the decrease in liquid flow velocity at the impeller outlet. As such these figures clearly illustrate the correlation between variations in liquid flow velocity and static pressure. Moreover, Figures 14–16 illustrate the impact of variations in blade height on static pressure, liquid flow velocity, and streamlined liquid velocity, respectively, where average liquid flow velocities at the impeller outlet were 15.92 m/s, 12.64 m/s, and 10.56 m/s for b_2 = 12 mm, b_2 = 18 mm, and b_2 = 24 mm, respectively.

Static pressure (Pa)

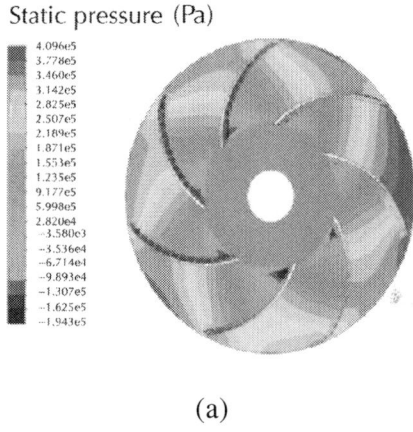

(a)

Static pressure (Pa)

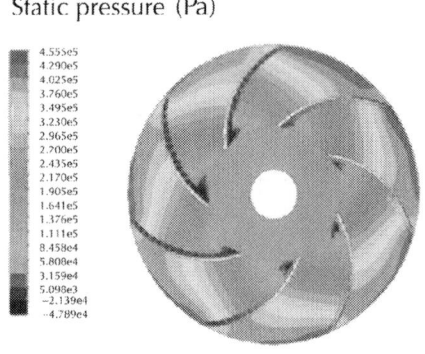

(b)

Static pressure (Pa)

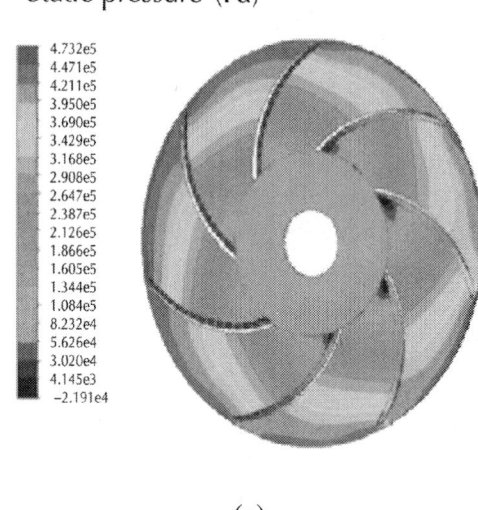

(c)

Figure 14: Static pressure contour.

Velocity (m/s)

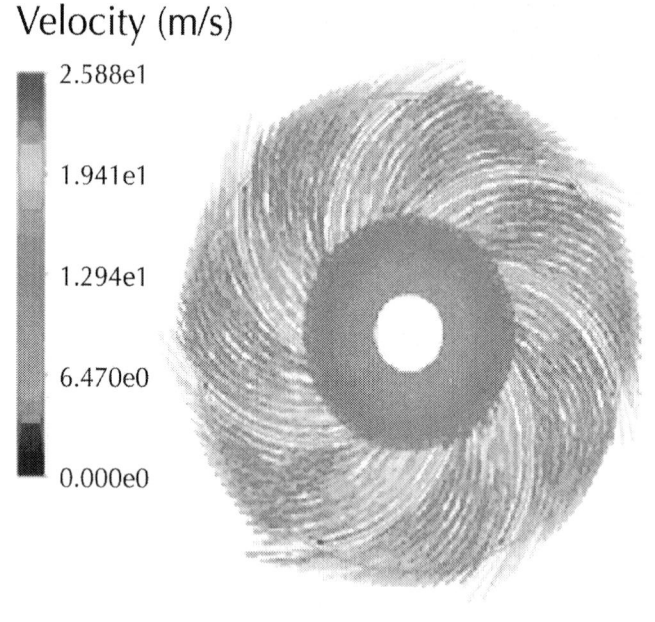

(a)

Velocity (m/s)

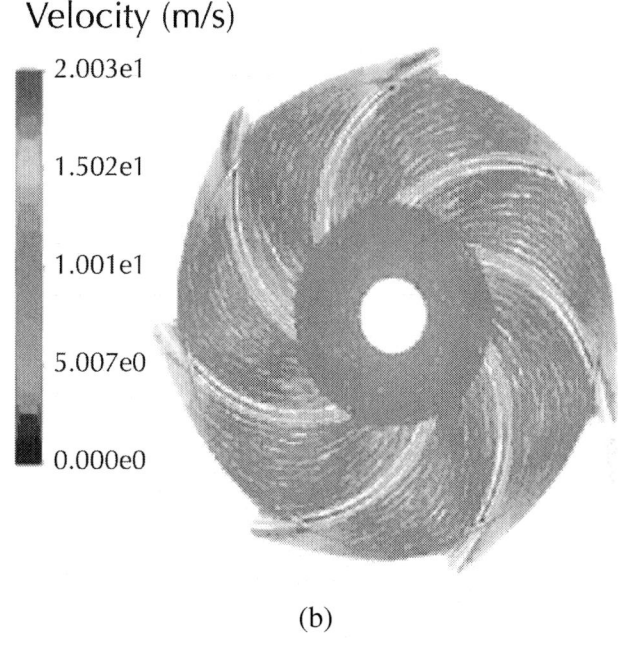

(b)

Velocity (m/s)

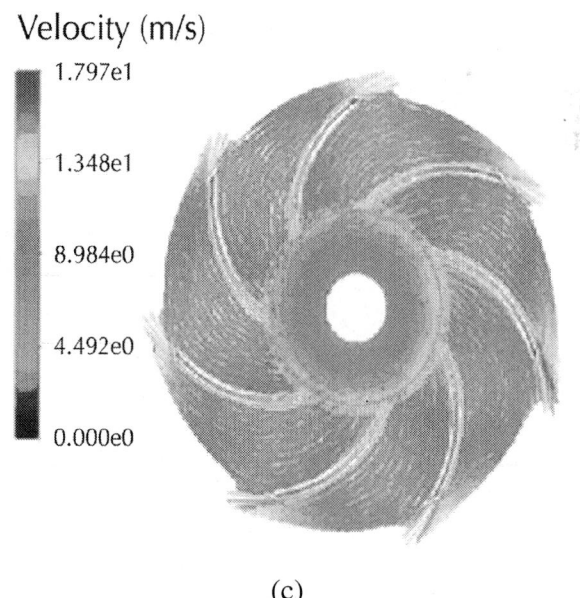

(c)

Figure 15: Liquid flow velocity vector.

Streamlines

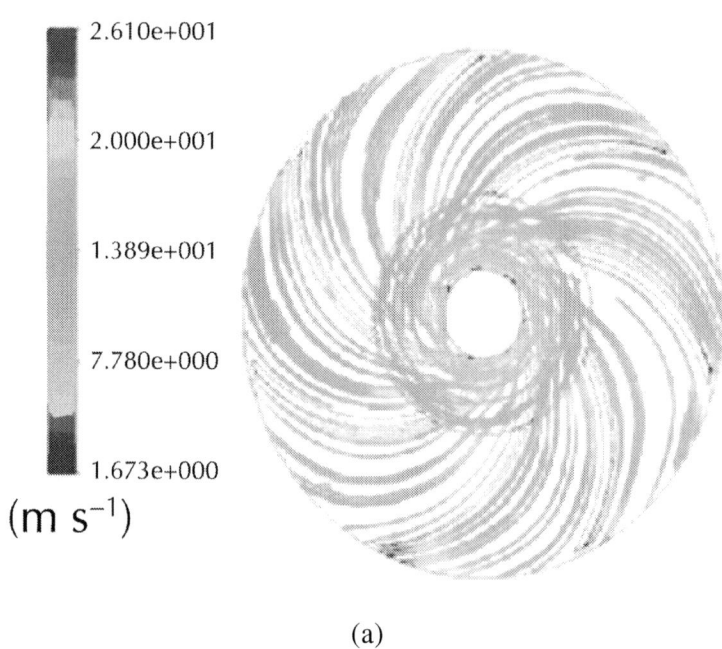

(a)

Streamlines

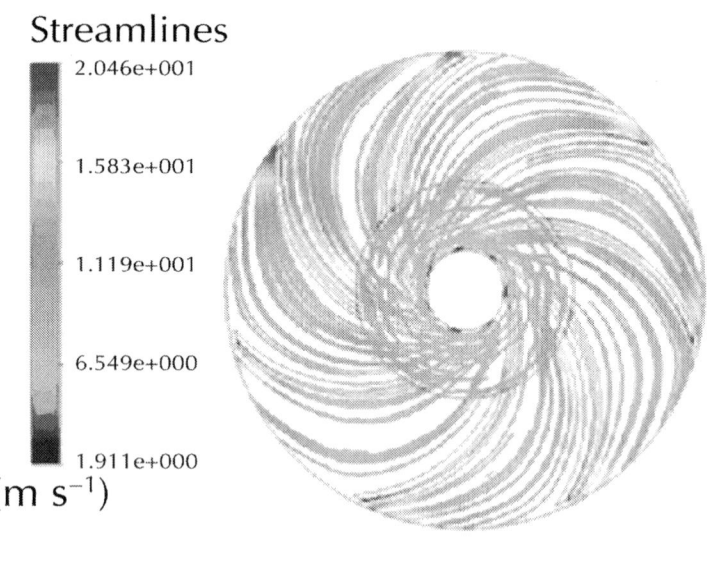

(b)

Streamlines

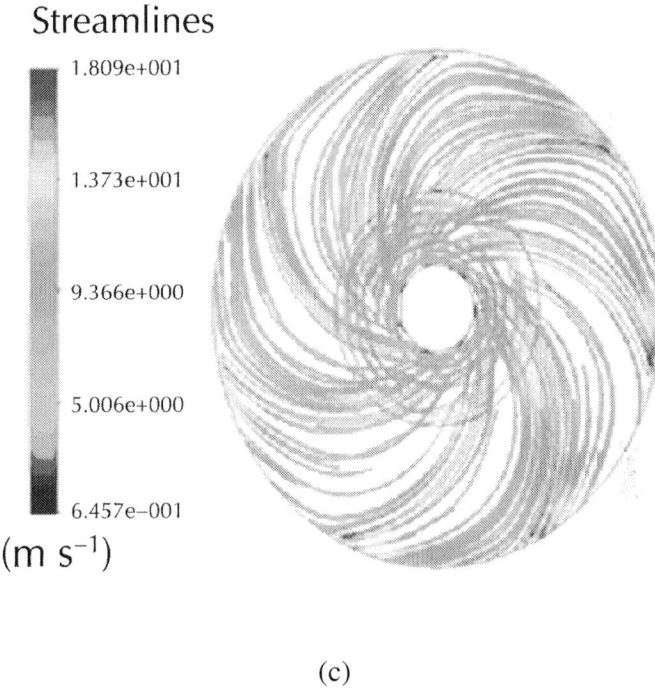

$$(m\ s^{-1})$$

(c)

Figure 16: Streamlines of liquid flow velocity.

Effect of Impeller Blade Width

To investigate the effect that the impeller blade width has on the pump head, the pump brake horsepower, and the pump overall efficiency, the blade widths of 4 mm, 10 mm, and 15 mm were selected, while the other parameters were keep constant. Figure 17 shows the pump head as a function of the volume flow rate, illustrating that the pump head decreases with increased blade width. This is due to augmenting the liquid pressure drop with increasing blade width. Also, the required pump brake horsepower decreases when the blade width rises, as indicated in Figure 18. The corresponding overall efficiency curves are shown in Figure19, illustrating that the blade width's impact on the overall efficiency is more pronounced in at high volume flow rates. In other words, the overall efficiencies for the three blade widths decrease rapidly to the right side of the BEP and the lowest overall efficiency is obtained when e = 15 mm.

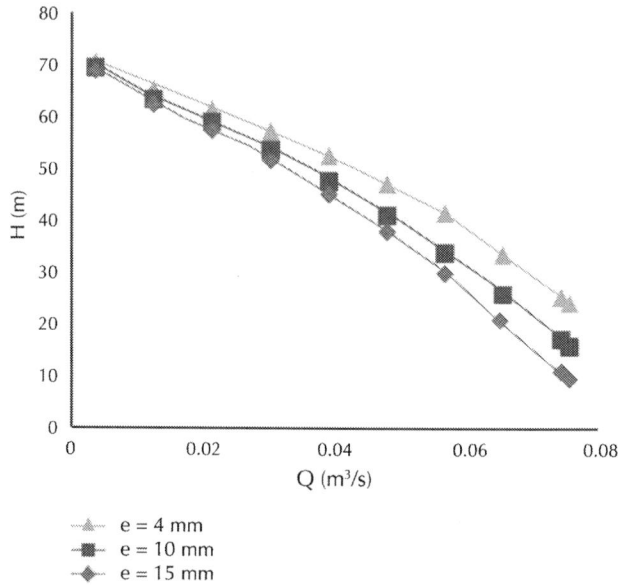

Figure 17: Pump head versus volume flow rate (parameter: blade width).

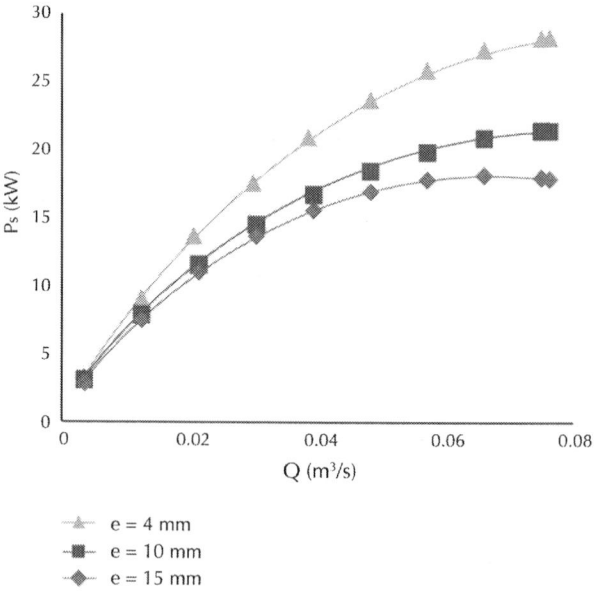

Figure 18: Pump brake horsepower versus volume flow rate (parameter: blade width).

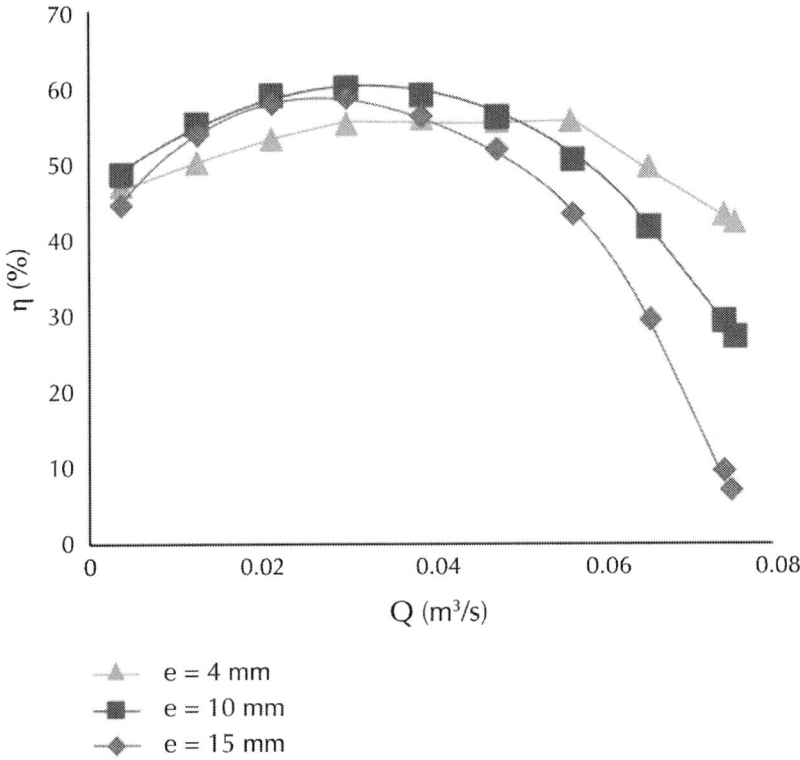

Figure 19: Overall efficiency versus volume flow rate (parameter: blade width).

Effect of Impeller Outlet Blade Angle

Three impeller outlet blade angle values of 9°, 28°, and 60° were selected to investigate their influence on the pump head, the pump brake horsepower, and the pump's overall efficiency. Figure 20 depicts the distribution of the pump head as a function of volume flow rate and with outlet blade angle as a parameter. This figure thus shows that the pump head increases with increasing outlet blade angle, which can be explained by the increased outlet cross-section size relative to the increased outlet blade angle, thus leading to diminution of liquid pressure drop in flow passage between blades.

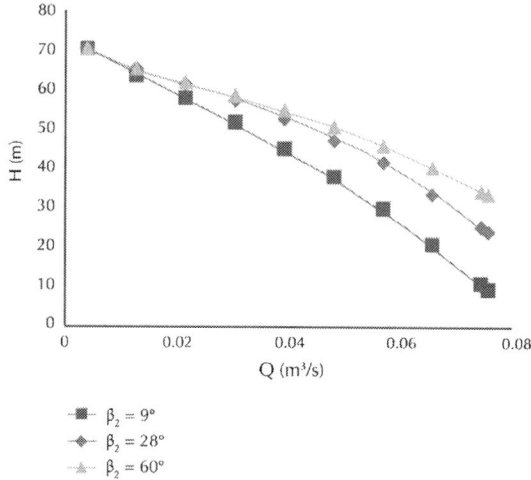

Figure 20: Pump head versus volume flow rate (parameter: outlet blade angle).

In addition, Figure 21 depicts the corresponding brake horsepower curves as a function of the volume flow rate, illustrating that the pump brake horsepower increases relative to the augmenting outlet blade angle. This is due to the increase in the requested shaft torque, along with the augmented outlet blade angle.

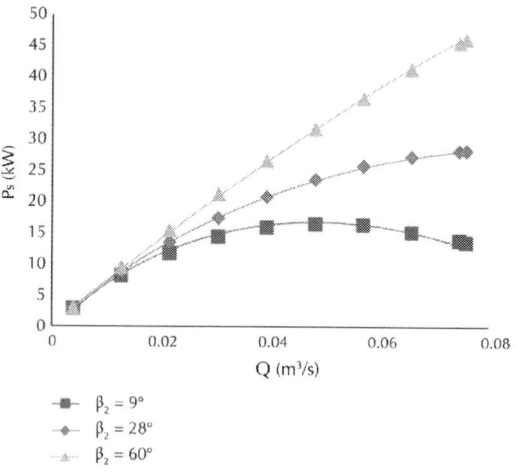

Figure 21: Brake horsepower versus volume flow rate (parameter: outlet blade angle).

Moreover, the efficiency curves shown in Figure 22 illustrate that the overall efficiency for $\beta_2=9°$ decreases rapidly to the right of the BEP.

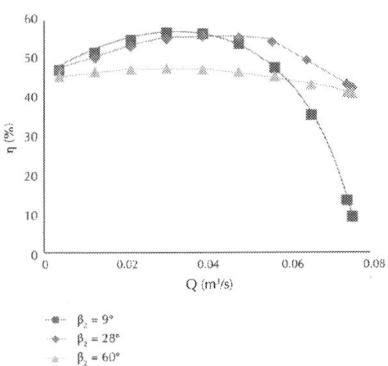

Figure 22: Overall efficiency versus volume flow rate (parameter: outlet blade angle).

Additionally, Figures 23 and 24 show the static pressure contour and the liquid flow velocity vector for $Q = 0.065 \ m^3/s$. From these figures, it can be observed that the static pressure difference between the impeller outlet and inlet increases with the augmented blade angle, due to a decrease in the liquid flow velocity at the impeller outlet, as indicated in Figure 26. The average liquid flow velocities at the impeller outlet are 21.06 m/s, 15.92 m/s, and 10.09 m/s for $\beta_2=9°, \beta_2=28°$, and $\beta_2=60°$, respectively.

Static pressure (Pa)

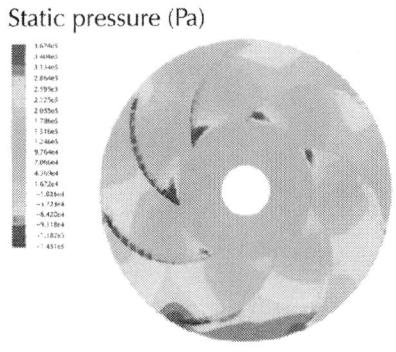

(a)

Static pressure (Pa)

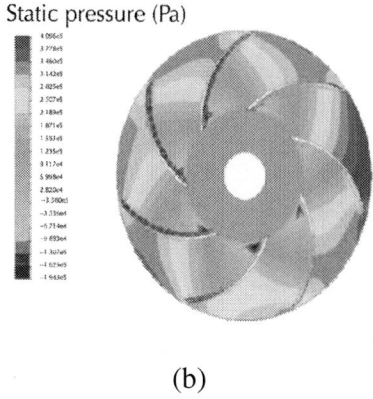

(b)

Static pressure (Pa)

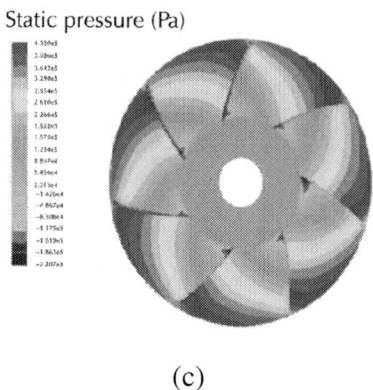

(c)

Figure 23: Static pressure contour.

Velocity (m/s)

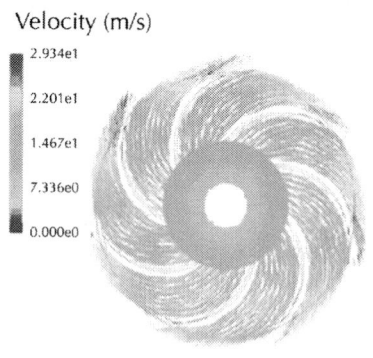

(a)

Velocity (m/s)

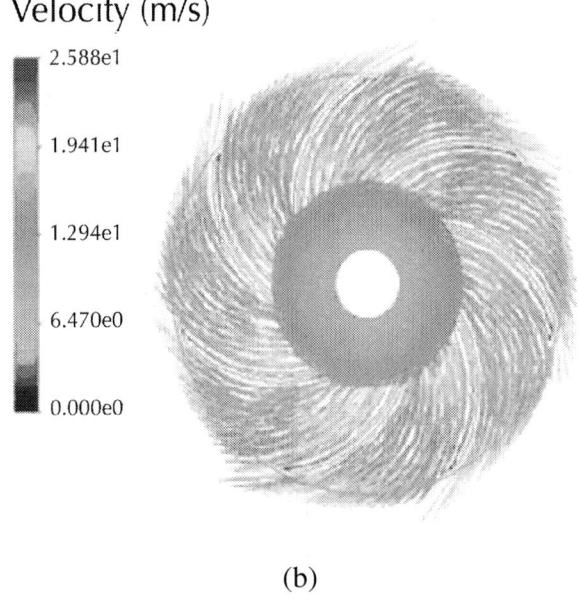

2.588e1

1.941e1

1.294e1

6.470e0

0.000e0

(b)

Velocity (m/s)

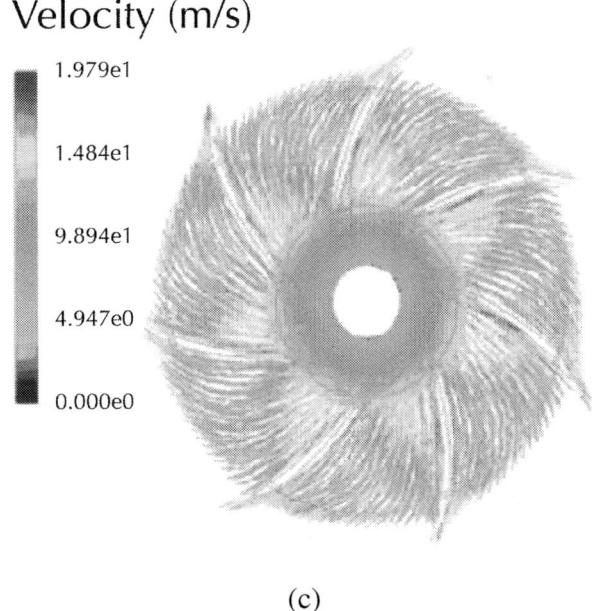

1.979e1

1.484e1

9.894e1

4.947e0

0.000e0

(c)

Figure 24: Liquid flow velocity vector.

Effect of Impeller Blade Number When Accounting for Volute

To investigate the effect of the impeller blade number on the pump head, the pump brake horsepower and the overall pump efficiency, three impellers whose blade number were 5, 7, and 9 were selected, while the other parameters were kept constant. Figure 25 shows the pump head as a function of the volume flow rate, illustrating that the pump head increases with a greater blade number. This is explained by the decrease in the liquid pressure drop in the flow passage with an augmented impeller blade number, keeping the same total volume flow rate. Also, as shown in Figure 26, the pump brake horsepower increases relatively with the augmented blade number. This is due to the increase in the request pump shaft torque, as the pump blade number also increases.

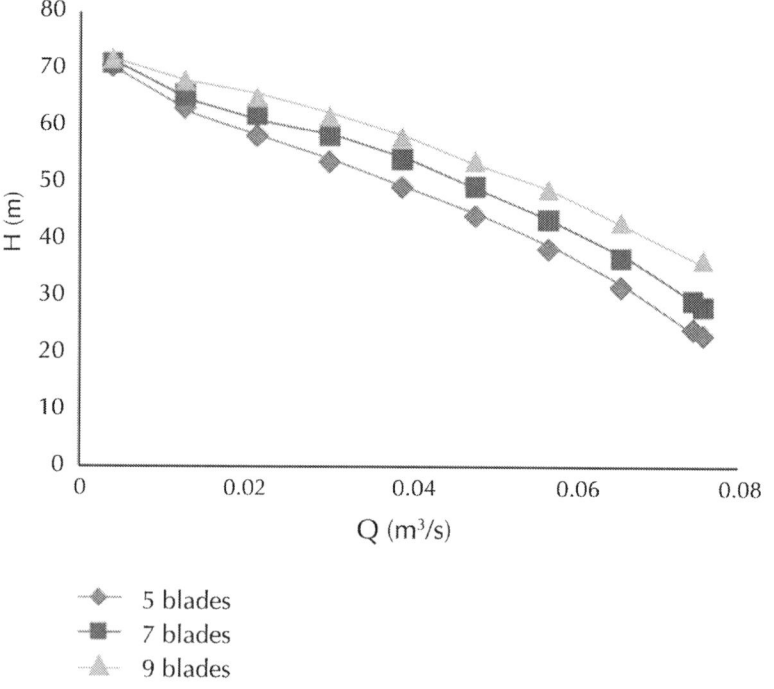

Figure 25: Pump head versus volume flow rate (parameter: impeller blade number).

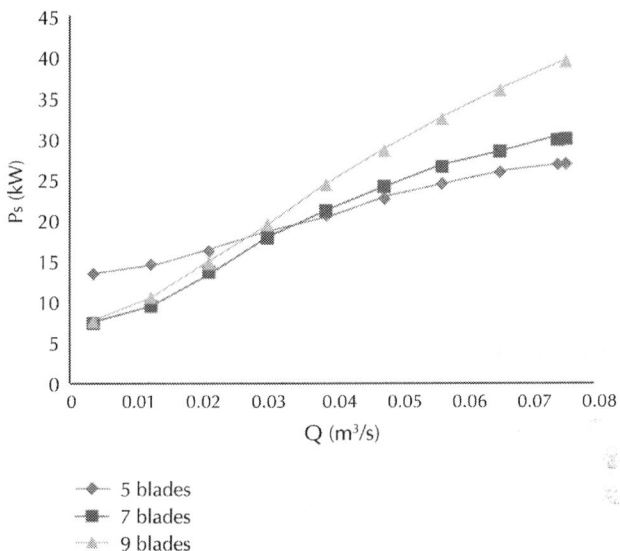

Figure 26: Brake horsepower versus volume flow rate (parameter: impeller blade number).

In addition, Figure 27 shows the overall efficiency curves, showing that the impeller having 5 blades has the lowest overall efficiency.

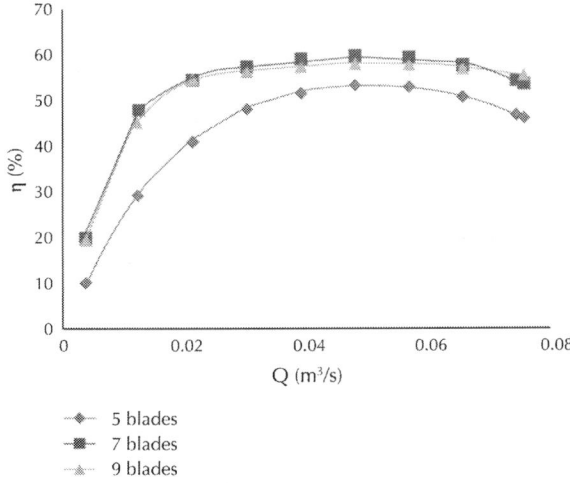

Figure 27: Overall efficient versus blade number (parameter: blade number).

Moreover, Figures 28 and 29 depict the corresponding static pressure contour and liquid flow velocity vector for Q = 0.065 m³/s, respectively. These figures thus clearly show the increased static pressure difference between the volute outlet and the impeller inlet relative to the increasing blade number. This confirms the reduction in the liquid flow velocity at the impeller outlet relative to the greater blade number, as represented in Figure 29, where the average liquid flow velocities at the impeller outlet were 16.06 m/s, 15.40 m/s, and 12.53 m/s for 5 blades, 7 blades, and 9 blades, respectively.

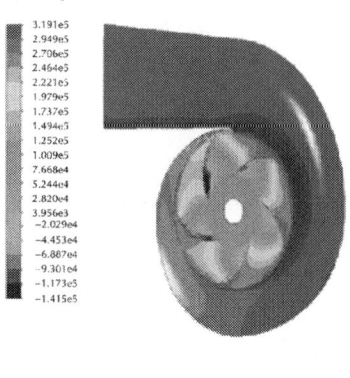

(a)

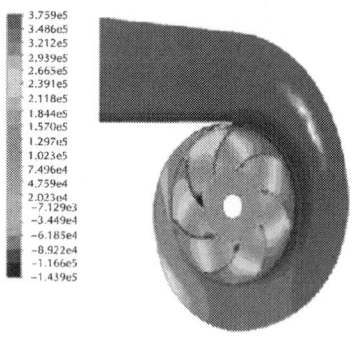

(b)

Static pressure (Pa)

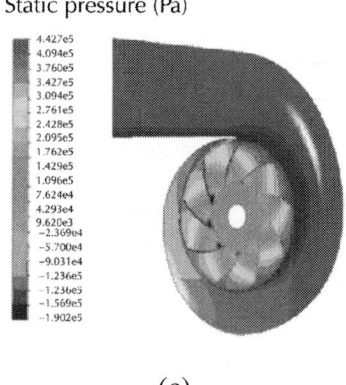

(c)

Figure 28: Static pressure contour.

Velocity (m/s)

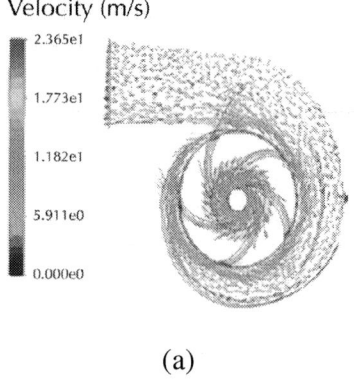

(a)

Velocity (m/s)

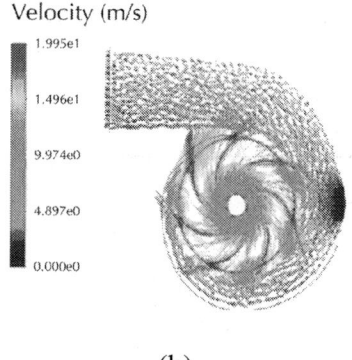

(b)

Velocity (m/s)

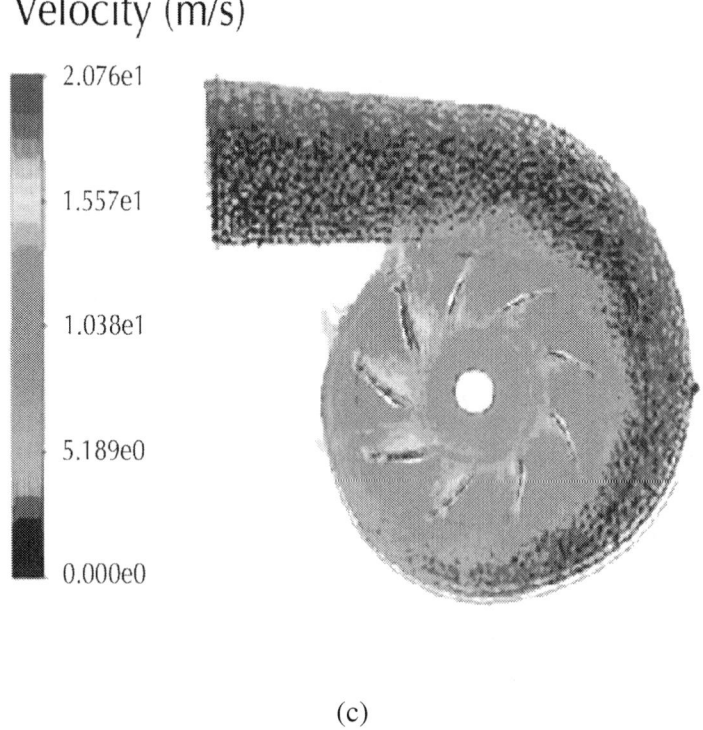

(c)

Figure 29: Vectors of liquid flow velocity contour.

Effect of Impeller Diameter When Accounting for Volute

The impeller outlet diameter values of 285 mm and 320 mm were selected to investigate their effects on pump performance when keeping the other parameters constant. Figure 30 shows that the pump head increases with increasing impeller diameter, which can be explained by the fact that the liquid static pressure drop in impeller decreases with increasing impeller diameter. In other words, for a given volume flow rate, the pressure difference between the volute outlet and the impeller inlet is higher for an impeller with a greater diameter. In addition, Figure 31 shows that the brake horsepower increases relative to the increasing impeller diameter, due to the requested augmented impeller shaft torque relative to the size of the impeller diameter.

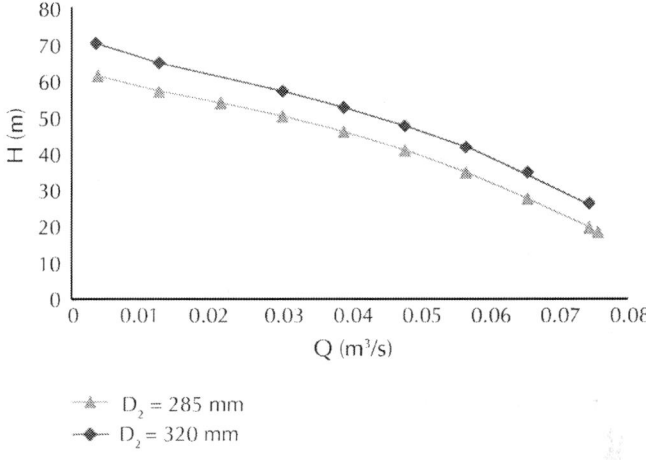

Figure 30: Pump head versus volume flow rate (parameter: impeller diameter).

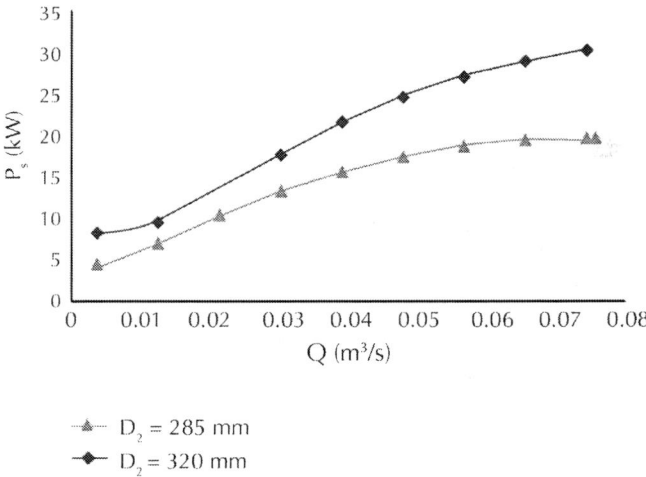

Figure 31: Brake horsepower versus volume flow rate (parameter: impeller diameter).

Moreover, the corresponding overall efficiency curves shown in Figure 32 indicate that the impeller having a great diameter has better overall efficiency with volume flow rates greater than 0.02 m³/s.

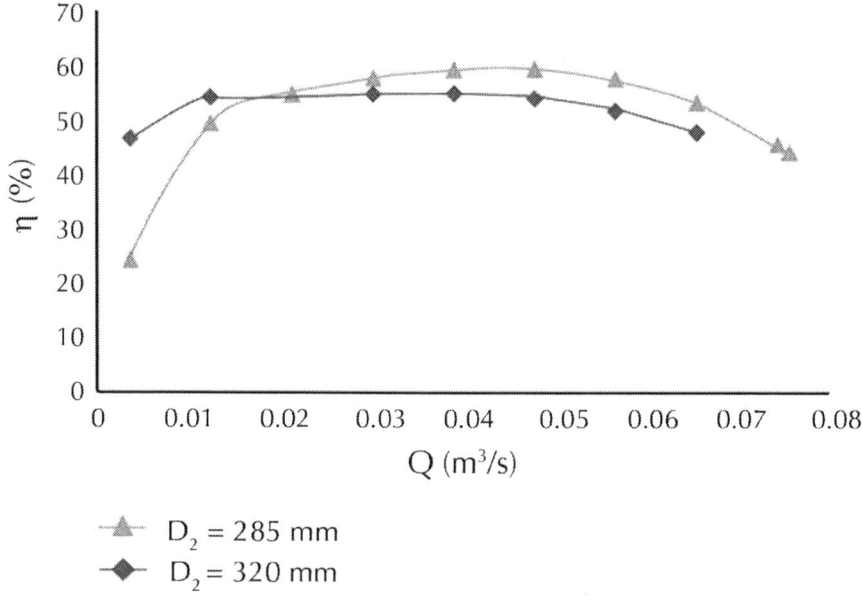

Figure 32: Overall efficiency versus volume flow rate (parameter: impeller diameter).

Effect of Diffuser Blade Number

To analyze the effect the diffuser blade number has on the pump head, the pump brake horsepower, and the overall pump efficiency, three diffuser models with blade numbers of 5, 8, and 12 were selected, while the other parameters were kept constant. Figure 33 shows the pump head as a function of the volume flow rate, where it is observed that the impact of the diffuser number on the pump head is small, even if the pump head for the diffuser blade number of 8 is highest for the Q between 0.012 m³/s and 0.055 m³/s. As depicted in Figure 34, the variation in brake horsepower due to diffuser blade number is also small, even if the diffuser blade number of 12 corresponds to a lowest brake horsepower.

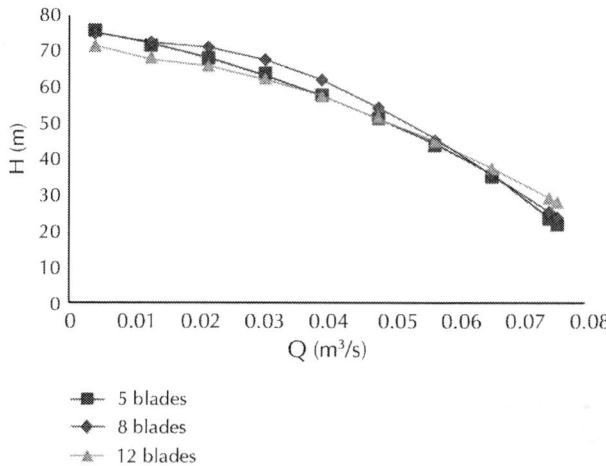

Figure 33: Pump head versus volume flow rate (parameter: diffuser blade number).

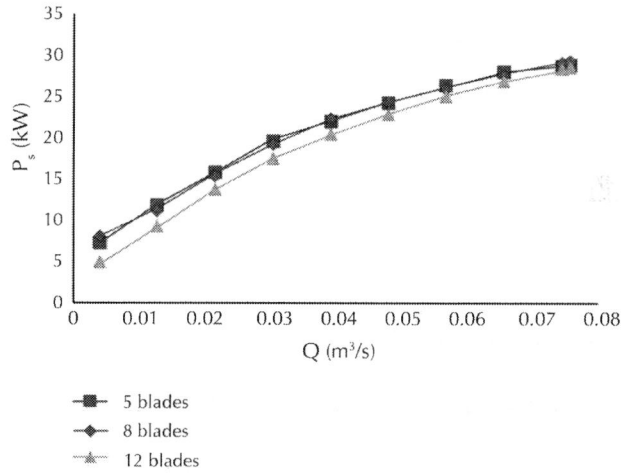

Figure 34: Brake horsepower versus volume flow rate (parameter: diffuser blade number).

Furthermore, Figure 35 shows that, for the low and the high volume flow rates, the overall efficiency for the diffuser blade number 12 is highest whereas the overall efficiencies for diffuser blade numbers of 5 and 8 are nearly the same. This figure also indicates that the overall efficiency is lowest for diffuser blade number 5

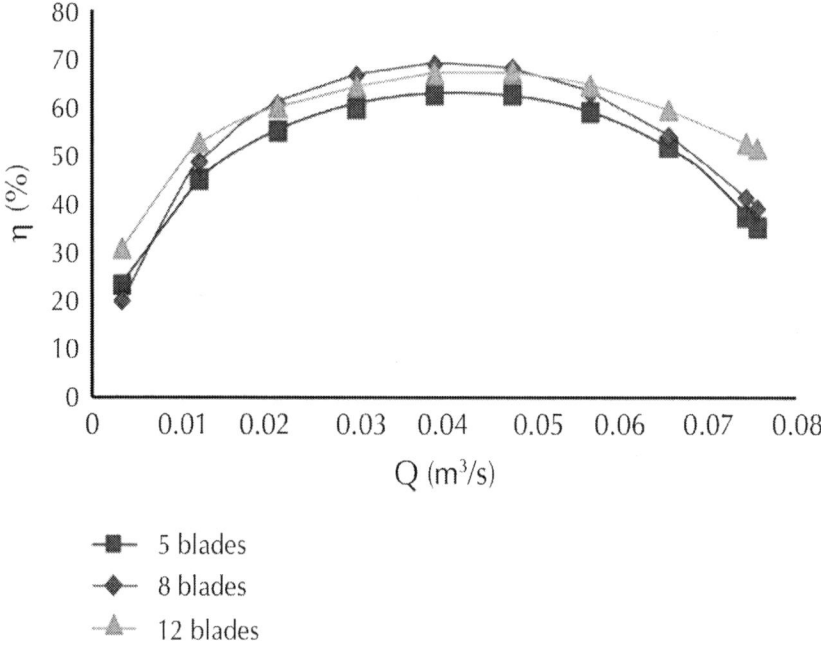

Figure 35: Overall efficiency versus volume flow rate (parameter: diffuser blade number).

Additionally, Figures 36 and 37 depict the corresponding static pressure contour and the liquid flow velocity vector for $Q = 0.065$, respectively, clearly showing for these figures the correlation between the increased static pressure difference and decreased liquid flow velocity at the diffuser outlet, with augmented diffuser blade number. Thus, the average values obtained for the static pressure difference between the diffuser outlet and the impeller inlet are 3.428×10^5 Pa, 3.49×10^5 Pa, and 3.65×10^5 Pa for blade numbers of 5, 8, and 12, respectively, as represented in Figure 36. Also, the average liquid flow velocity values at the diffuser outlet of 15.13 m/s, 12.22 m/s, and 9.06 m/s were found for the blade numbers of 5, 8, and 12, respectively, as shown in Figure 37.

Static pressure (Pa)

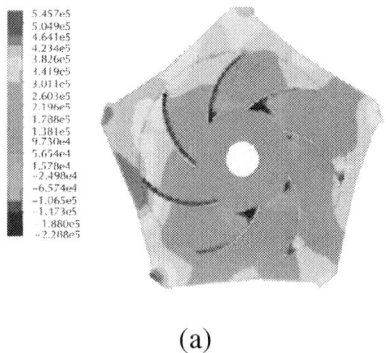

(a)

Static pressure (Pa)

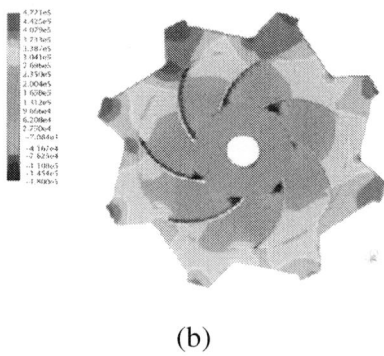

(b)

Static pressure (Pa)

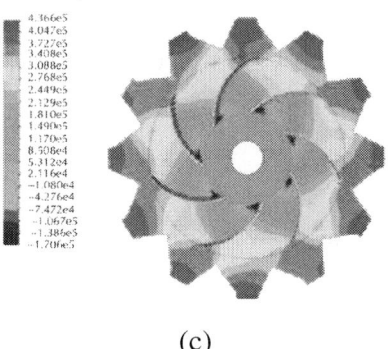

(c)

Figure 36: Static pressure contour.

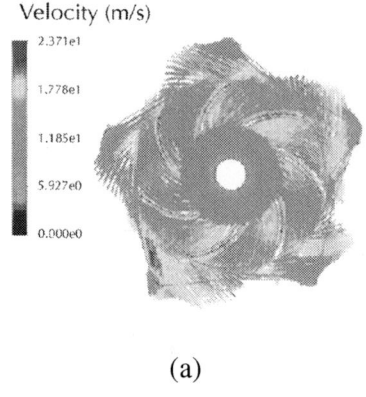

(a)

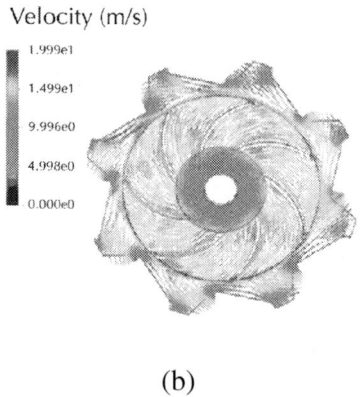

(b)

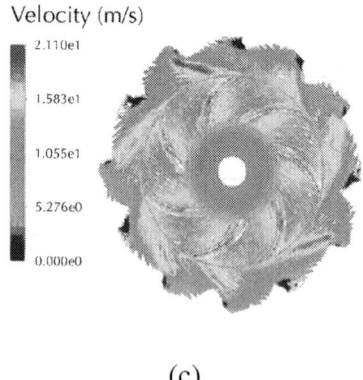

(c)

Figure 37: Vectors of liquid flow velocity.

Model Comparison

The impeller and diffuser combination was selected to validate the numerical approach developed, since the experimental results of this case were available from Technosub Inc. When accounting for experimental boundary conditions for the numerical simulations run, Figures 38–40 show the comparison between the experimental and the numerical results for the pump head, the brake horsepower, and the overall efficiency. The discrepancies observed could be explained by the fact that lost mechanical power, power lost due to leakage, and the pump casing were not taken into account in the numerical simulations carried out. The horsepower for experimental pump brake was therefore higher than the numerical brake horsepower obtained, as illustrated in Figure 39.

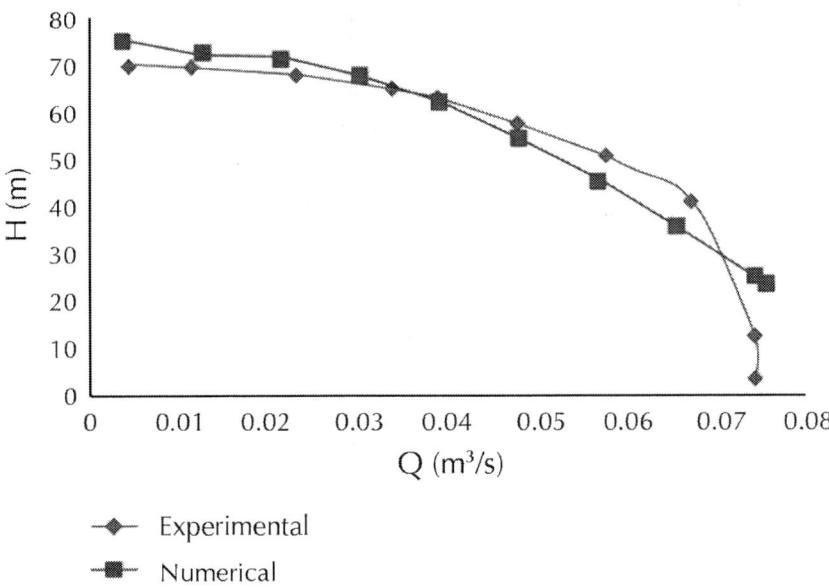

Figure 38: Pump head versus volume flow rate (parameter: numerical and experimental results).

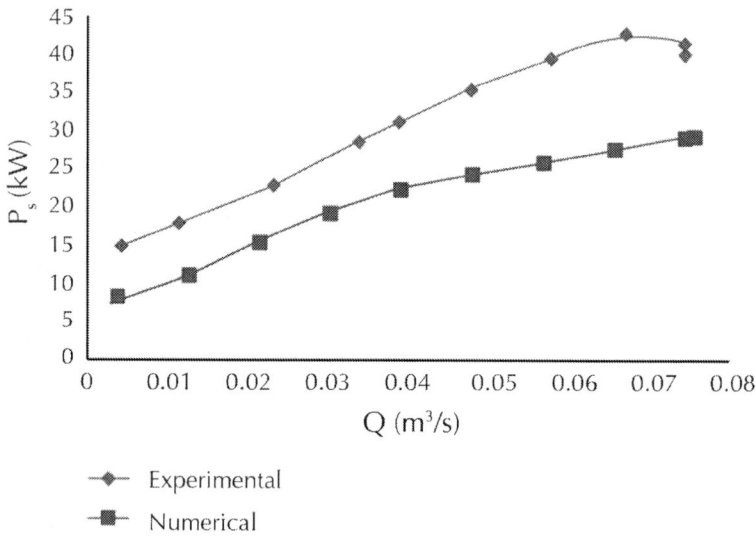

Figure 39: Brake horsepower versus volume flow rate (parameter: numerical and experimental results).

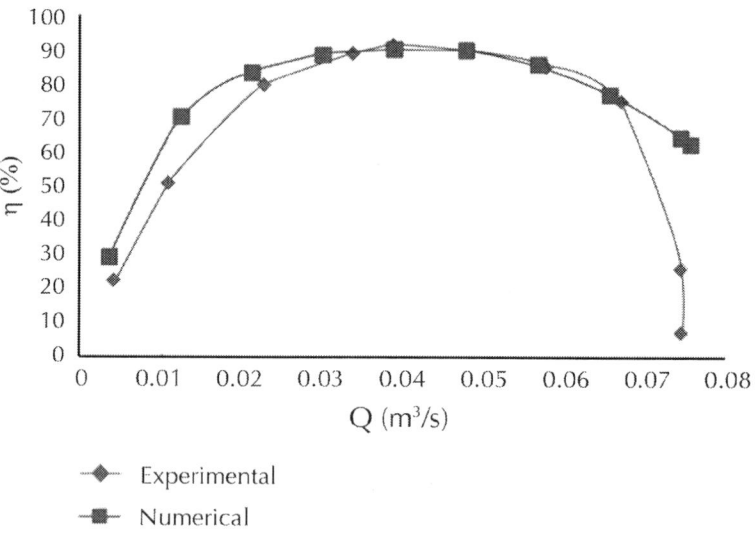

Figure 40: Overall efficiency versus volume flow rate (parameter: numerical and experimental results).

CONCLUSIONS

In this study, a steady-state liquid flow in a three-dimensional centrifugal pump was numerically investigated. Models of impeller, combined impeller and volute, and combined impeller and diffuser were developed to analyze the effects the key design parameters, including the blade height, the outlet blade angle, the blade width, the blade number, and the impeller outer diameter, had on the pump head, the brake horse power, and the overall efficiency. The obtained results demonstrate, among others, that the pump head and the brake horsepower increase with increasing impeller blade number and impeller blade height, while they decrease with increasing impeller blade width. Also, the interaction between the impeller and the volute reveals that the decrease of the impeller outer diameter keeping the volute dimensions constant leads to the reduction of the pump head and the brake horsepower. The pump overall efficiency is also influenced by the selected key design parameter. A relatively good agreement was observed comparing the developed numerical approach with the experimental results for the case of the combined impeller and diffuser obtained from a pump manufacturer.

ACKNOWLEDGMENTS

The authors are grateful to the Foundation of University of Quebec in Abitibi-Temiscamingue (FUQAT) and the company Technosub Inc.

REFERENCES

1.   W. W. Peng, Fundamentals of Turbomachinery, John Wiley and Sons, Hoboken, NJ, USA, 2008.

2.   W. Zhou, Z. Zhao, T. S. Lee, and S. H. Winoto, "Investigation of flow through centrifugal pump impellers using computational fluid dynamics," International Journal of Rotating Machinery, vol. 9, no. 1, pp. 49–61, 2003.

3.   J. S. Anagnostopoulos, "A fast numerical method for flow analysis and blade design in centrifugal pump impellers," Computers & Fluids, vol. 38, no. 2, pp. 284–289, 2009.

4.   B. Cui, Z. Zhu, J. Zhang, and Y. Chen, "Flow simulation and experimental study of low-specific-speed high-speed complex centrifugal impellers," Chinese Journal of Chemical Engineering, vol. 14, no. 4, pp. 435–441, 2006.

5.   S. Derakhshan and A. Nourbakhsh, "Theoretical, numerical and experimental investigation of centrifugal pumps in reverse operation," Experimental Thermal and Fluid Science, vol. 32, no. 8, pp. 1620–1627, 2008.

6.   R. Spence and J. Amaral-Teixeira, "Investigation into pressure pulsations in a centrifugal pump using numerical methods supported by industrial tests," Computers & Fluids, vol. 37, no. 6, pp. 690–704, 2008.

7.   K. W. Cheah, T. S. Lee, S. H. Winoto, and Z. M. Zhao, "Numerical flow simulation in a centrifugal pump at design and off-design conditions," International Journal of Rotating Machinery, vol. 2007, Article ID 83641, 8 pages, 2007. ·

8.   M. H. Shojaee Fard, F. A. Boyaghchi, and M. B. Ehghaghi, "Experimental study and three-dimensional numerical flow simulation in a centrifugal pump when handling viscous fluids," International Journal of Engineering Science, vol. 17, no. 3-4, pp. 53–60, 2006.

9.   R. K. Byskov, C. B. Jacobsen, and N. Pedersen, "Flow in a centrifugal pump impeller at design and off-design conditions—part II: large eddy simulations," Journal of Fluids Engineering, vol. 125, no. 1, pp. 73–83, 2003.

10.   K. Majidi, "Numerical study of unsteady flow in a centrifugal pump," Journal of Turbomachinery, vol. 127, no. 2, pp. 363–371, 2005.

11.   W.-G. Li, F.-Z. Su, and C. Xiao, "Influence of the number of impeller blades on the performance of centrifugal oil pumps," World Pumps, vol. 2002, no. 427, pp. 32–35, 2002.

12.   H. Liu, Y. Wang, S. Yuan, M. Tan, and K. Wang, "Effects of blade number on characteristics of centrifugal pumps," Chinese Journal of Mechanical Engineering, vol. 23, no. 6, pp. 742–747, 2010. ·

13.   M. H. Shojaee Fard and F. A. Boyaghchi, "Studies on the influence of various blade outlet angles in a centrifugal pump when handling viscous fluids," American Journal of Applied Sciences, vol. 4, no. 9, pp. 718–724, 2007.

14. E. C. Bacharoudis, A. E. Filios, M. D. Mentzos, and D. P. Margaris, "Parametric study of a centrifugal pump impeller by varying the outlet blade angle," The Open Mechanical Engineering Journal, vol. 2, pp. 75–83, 2008.

15. M. Gölcü and Y. Pancar, "Investigation of performance characteristics in a pump impeller with low blade discharge angle," World Pumps, vol. 2005, no. 468, pp. 32–40, 2005. ·

16. J. González, J. Fernández-Francos, E. Blanco, and C. Santolaria-Morros, "Numerical simulation of the dynamic effects due to impeller-volute interaction in a centrifugal pump," Journal of Fluids Engineering, vol. 124, no. 2, pp. 348–355, 2002.

17. M. Asuaje, F. Bakir, S. Kouidri, F. Kenyery, and R. Rey, "Numerical modelization of the flow in centrifugal pump: volute influence in velocity and pressure fields," International Journal of Rotating Machinery, vol. 2005, no. 3, pp. 244–255, 2005.

18. K. A. Kaupert and T. Staubli, "The unsteady pressure field in a high specific speed centrifugal pump impeller—part I: influence of the volute," Journal of Fluids Engineering, vol. 121, no. 3, pp. 621–626, 1999.

19. P. Hergt, S. Meschkat, and B. Stoffel, "The flow and head distribution within the volute of a centrifugal pump in comparison with the characteristics of the impeller without casing," Journal of Computational and Applied Mechanics, vol. 5, no. 2, pp. 275–285, 2004.

20. J. González Pérez, J. Parrondo, C. Santolaria, and E. Blanco, "Steady and unsteady radial forces for a centrifugal pump with impeller to tongue gap variation," Journal of Fluids Engineering, vol. 128, no. 3, pp. 454–462, 2006.

21. A. Ozturk, K. Aydin, B. Sahin, and A. Pinarbasi, "Effect of impeller-diffuser radial gap ratio in a centrifugal pump," Journal of Scientific and Industrial Research, vol. 68, no. 3, pp. 203–213, 2009. Ansys inc., ANSYS-CFX, User Manual, USA, 2008.

22. Technosub Inc., http://technosub.net/.

23. G. Lemasson, "Les machines transformatrices d›énergie," TOME II, Turbomachine-Machine alternatives, 1967.

Chapter 5

A New Proposed Return Guide Vane for Compact Multistage Centrifugal Pumps

Qihua Zhang[1], Weidong Shi[1], Yan Xu[1], Xiongfa Gao[1], Chuan Wang[1], Weigang Lu[1], and Dongqi Ma[2]

[1]National Research Center of Pumps and Pumping System Engineering and Technology, Jiangsu University, Zhenjiang 212013, China

[2]Fujian Academy of Mechanical Sciences, Fuzhou, Fujian 350005, China

ABSTRACT

For widely used multistage centrifugal pumps, their former structures are so bulky that nowadays growing interest has been shifted to the development of more compact structures. Following this trend, a compact pump structure is provided and analysed. To maintain the pump's pressure recovery, as well as to meet the water flow from the impeller, a circumferential twisted return guide vane (RGV) is proposed. To validate this design method, the instantaneous CFD simulations are performed to investigate the rotor-stator interventions. Within the impeller, the pressure fluctuation is cyclic symmetry, where the impeller frequency dominates. At the zone where flow leaves impeller for RGV, the pressure fluctuation is nonperiodic, the impeller frequency is major, and the rotation frequency is secondary. Within RGV, the periodic symmetric fluctuation is recovered, where the rotation frequency is governing. The fluctuation decreases from seven cycles within impeller to two cycles within RGV, indicating that the flow from impeller is well handled by RGV. To examine the pump's performance, a prototype multistage pump is designed. The testing shows that the pump efficiency is 57.5%, and the stage head is 9 m, which is comparable to former multistage centrifugal pumps. And this design is more advantageous in developing compact multistage centrifugal pumps.

INTRODUCTION

To meet variable flow angles into gas turbines, hydraulic runners, and so forth, inlet guide vane (IGV) is adopted to keep them operating at peak performance [1–5]. For compressors, fans, pumps, and so forth, IGV is also used to manipulate their operating load [6–11]. Outlet guide vane (OGV) is heavily used at downstream of low pressure turbines, fans, compressors, and pumps, where the rotating velocity can be effectively transformed to static pressure, reducing flow-induced vibration and noise [12–19]. Nozzle guide vane (NGV) is widely used in turbines. More specially, return guide vane (RGV) is applied for multistage centrifugal turbomachines, where RGV is functioned as an OGV with respect to its upstream stage; meanwhile, it is served as an IGV with respect to its downstream stage, which increases its design complexity [20–23].

In the first place, a suitable guide vane design is critical for overall stage performance. For IGV and OGV, flow in a mainstream direction is concerned, so airfoil is widely adopted in the streamwise direction, while in the spanwise direction, such strategies as free-vortex method, forced-vortex method, and radial-equilibrium method are available. To improve the performance of an impulse turbine, a free-vortex method was used to design a 3D IGV [1, 2], which achieved a 4.5% efficiency growth. And through a dual-curvature shape technique, an IGV of minihydraulic bulb turbines was proposed, which was validated by prototype turbine tests [3].

In early times, experimental testings were conducted to determine the effects of guide vanes on pump performance. And through comparisons of four guide vane arrangements in terms of pressure head, efficiency, and velocity distributions, it was concluded that RGV played a more crucial role in their multistage pump [20]. With technology advancement, especially with increasing CFD involvement since the 1980s, optimization has been popular in contemporary turbomachine design [24]. And especially for multistage turbomachines, the focus is the interaction between the rotor and the stator [25]. And sliding mesh technique was used to examine different flow patterns within a multistage pump by [26]. Moreover, rotor-stator interactions were investigated to aid in their design optimizations. By using a rotor-stator shear system, the drop size distribution of emulsification is numerically examined, and it was found that the effective equilibrium drop size was dominated by the rotor shape and its rotating speed [27]. Different combinations of flat type blade and curved-type blade in turbomolecular pumps were numerically investigated with DSMC, and the flat-curved rotor-stator combination was found to be more efficient [28]. The effect of the guide plate on the cavitation erosion on the bottom surface of the guide vane in three Gorges turbine was numerically investigated, and a new vortex structure clarifying this erosion was identified by the numerical simulation, which would be used to design antierosion guide plates [29].

On the other hand, optimal strategy research plays an important role in the turbomachine optimization design. A multiobjective method was performed to optimize a helico-axial pump impeller shape [30]. Monitoring and controlling strategy is an alternative procedure which can be calibrated to manage energy consumption of overall turbomachine systems. Commonly, this is an interdisciplinary and interdepartmental task. And a wide range of statistical investigations were performed upon a large number of pumps in plants to pursue energy saving [31]. And some monitoring techniques have been developed to improve existing pumping system efficiency [32].

Though multistage centrifugal pumps are not so widely applied as turbines, fans, they are indispensable in the deep well water and oil pumping and occupy a considerable market share of household water supply, rural drainage, and urban water circulation. So light weight and mobility are the most important factors concerned by the customers, challenging the former structure of multistage centrifugal pumps [33]. In this study, the difference between former pump structure and compact pump structure is firstly analyzed. And the merit of compact structure is addressed. Then, a new RGV design is proposed to match this compact structure. And the instantaneous rotor-stator interactions are conducted to evaluate RGV performance. Finally, a prototype multistage centrifugal pump is designed and tested to validate the design strategy.

RGV STRUCTURE STUDY

Due to inertia, fluid discharged from the impeller trailing edge is ongoing with high rotating speed. But in the ducted pumps, the circumferential velocity is useless. And RGV is commonly installed at downstream to recover this rotating velocity. There are two representative structures. The first type is widely used in boiler water feeding and power plant water circulations, where the stages are cascaded as depicted in Figure 1(a). The RGV blade structure is commonly designed to be cylindrical, as shown in Figures 1(b) and 1(c).

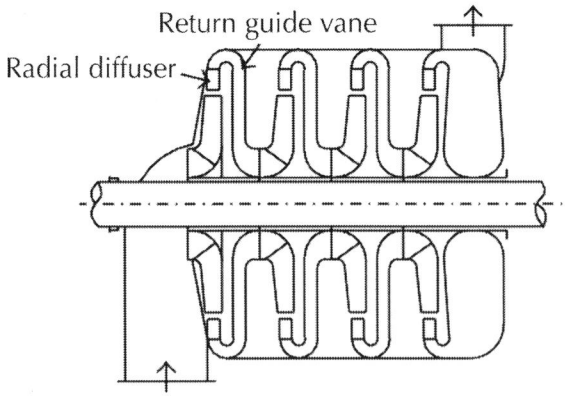

(a)

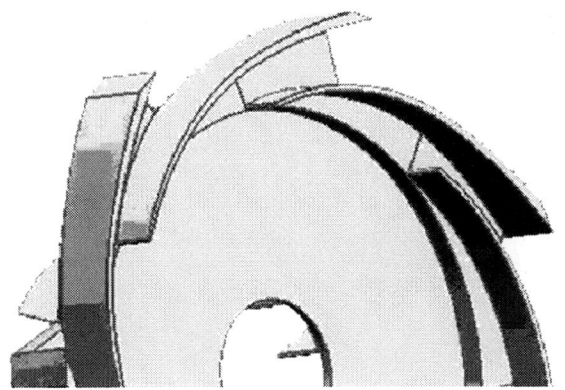

(b)

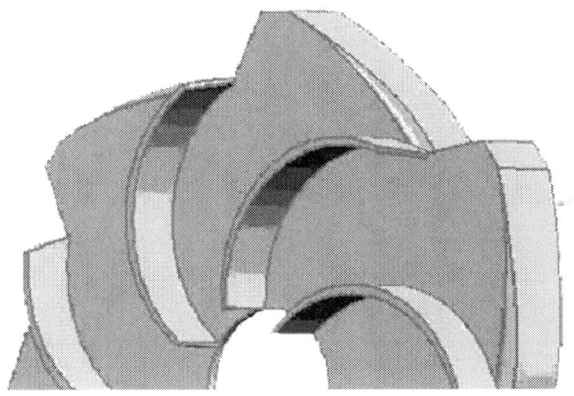

(c)

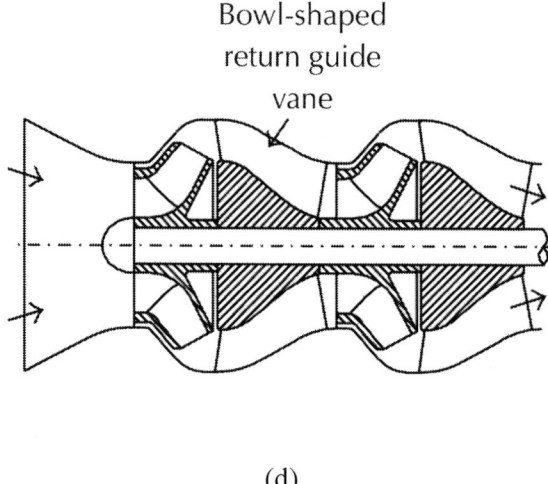

(d)

Figure 1: Two multistage pumps: (a) multistage pump equipped with radial diffuser and RGV, (b) radial diffuser, (c) RGV, (d) multistage pump equipped with bowl-shaped RGV, and (e) bowl-shaped RGV.

The second type is commonly used in oil drilling and rural and urban water supply from underground deep wells, so it is also called multistage submersible pump or the deep-well pump. As shown in Figure 1(e), the RGV blade structure is twisted like a bowl.

Structure Analysis

A compact stage for submersible pumps was proposed independently by [34, 35], where the impeller outer diameter is nearly equal to the casing diameter, increasing its stage head. To illustrate the difference between former structure and compact structure, a structure comparison is depicted in Figure 2. Herein, the compact stage means removing the radial diffuser; thus, it can be speculated that there are two extreme cases as depicted in Figures 2(b) and 2(c). For special case I, the impeller diameter D_{21} is not changed, and the casing diameter D_{31} shrinks to D_{32}, and its radial size is greatly reduced. For special case II, the casing diameter $D31$ is not changed, and the impeller diameter D_{21} is increased to D_{22}, and its head is theoretically multiplied by a factor $(D_{22}/D_{21})^2$; thus, less stages are needed and its axial size is shortened. So compact also means saving cost.

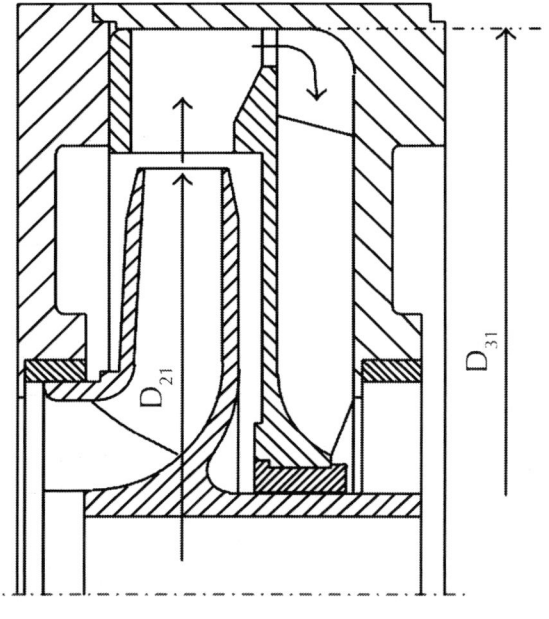

(a)

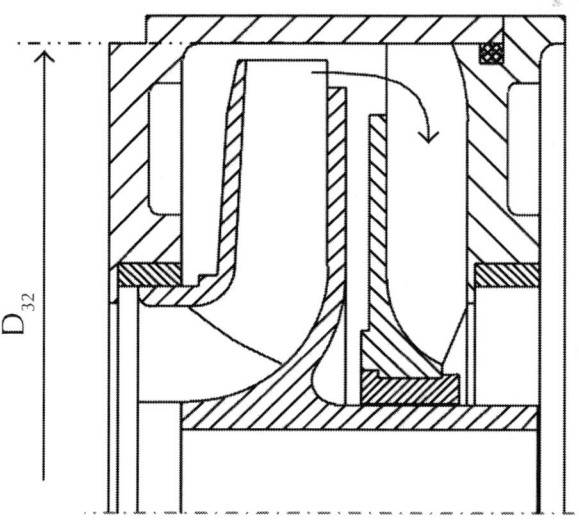

(b)

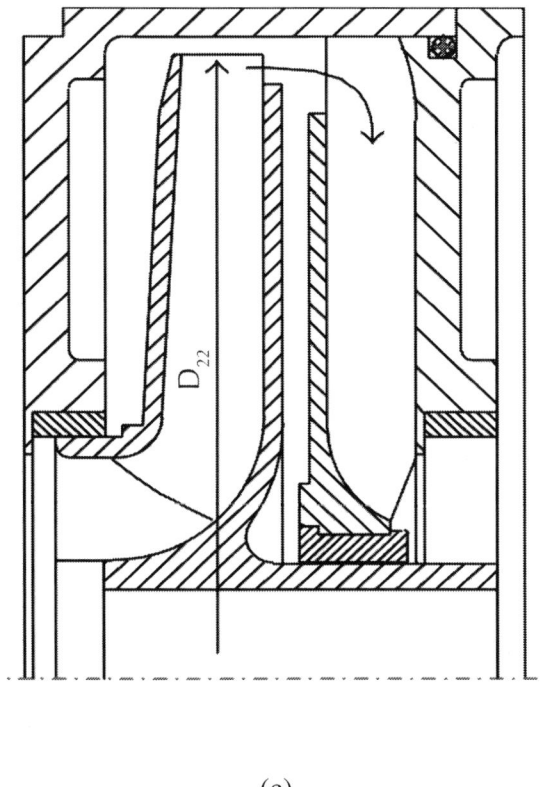

(c)

Figure 2: Structure comparison: (a) former structure, (b) special case I, and (c) special case II.

On the other hand, the special RGV also is an alternative to the bowl-shaped RGV from the cost saving aspect. But this is beyond our current research and would be regarded in the future.

Compact Stage Design

To extract oil or water from deep wells, the pump structure must be carefully tailored to utilize the limited space in well bore. In our early work [35], a compact deep-well pump was developed, as shown in Figure 3.

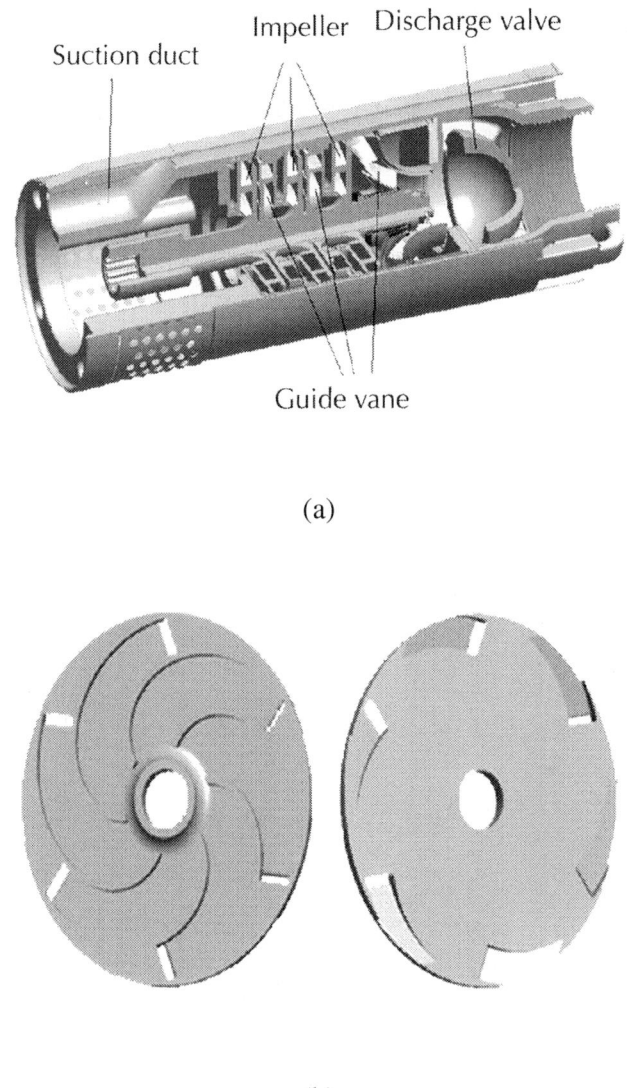

(a)

(b)

Figure 3: A multistage pump with compact stages: (a) assembly and (b) RGV.

This pump has only three stages, but it's performance exceeds the former multistage pump with five stages. To promote its popularity and to replace former multistage pumps, it is essential to establish a set of mature design procedures, especially for RGV. Though such efforts as steady-state simulation and experimental testing have been conducted to improve its hydraulic performance, cylindrical blade shape generally impedes through flow performance of RGV [36–39], as well as its pressure recovery. This point can be demonstrated by the principles of centrifugal pumps [40], where flow angles are varied along the leading edge to reduce the incidence loss. Therefore, a new circumferential twisted guide vane is proposed to tackle this problem.

A NEW RGV STRUCTURE

Twisted Blade Design Principle

A new RGV structure is developed by three streamlines at different spanwise position. As shown in Figure 4(a), by fixing the top streamline c, extending the mid span streamline b along the circumferential direction and further extending the bottom streamline a forward, a triple-streamline surface is constructed, where the three streamlines are drawn by our in-house code NQSJ [41], which is a general purpose RGV profile design tool. Then by cutting off the triangle block ABC from the round plate, a channel passage is formed, as shown in Figure 4(a).

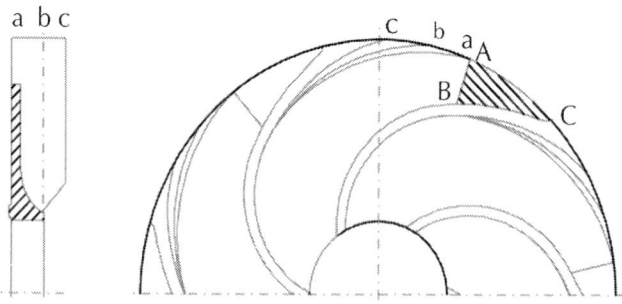

(a)

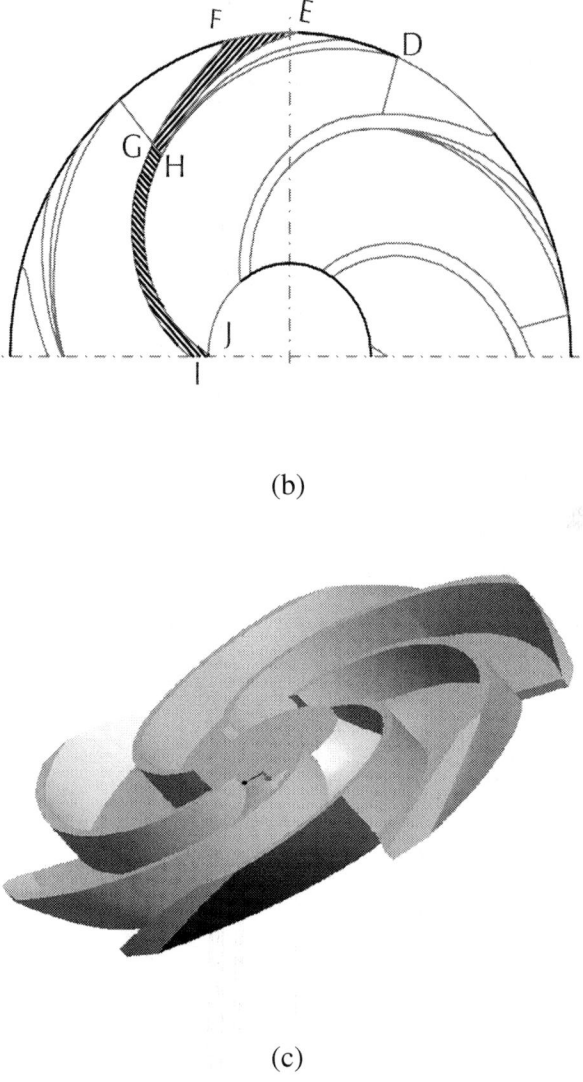

(b)

(c)

Figure 4: Blade design procedure: (a) schematic of a new RGV, (b) two parts of RGV, and (c) 3D view.

Further, as shown in Figure 4(b), the blade is combined by two joined parts: a cylindrical part $GHIJ$ and a twisted part $DEFGH$. Then, a practical RGV is formed as shown in Figure 4(c).Therefore, the key point is to establish the three streamlines a,b,c and the dividing point G, which will be discussed hereafter.

Twisted Blade Design Procedure

Herein, it is assumed that the impeller design has been finished. And our major concern is the geometric structure of the impeller discharge. To meet design flow rate, the impeller must be cut at the outer edge to achieve enough flow area. As shown in Figure 5, there are two cutting proposals. The first proposal cuts the blade as well as its hub plate, and the second proposal keeps the blade and cuts its hub plate.

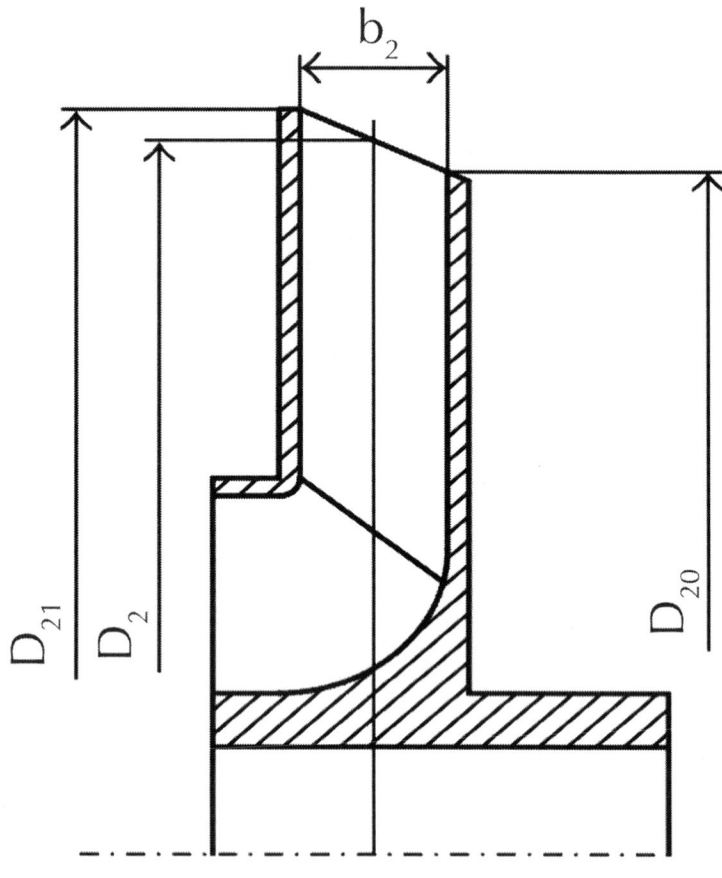

(a)

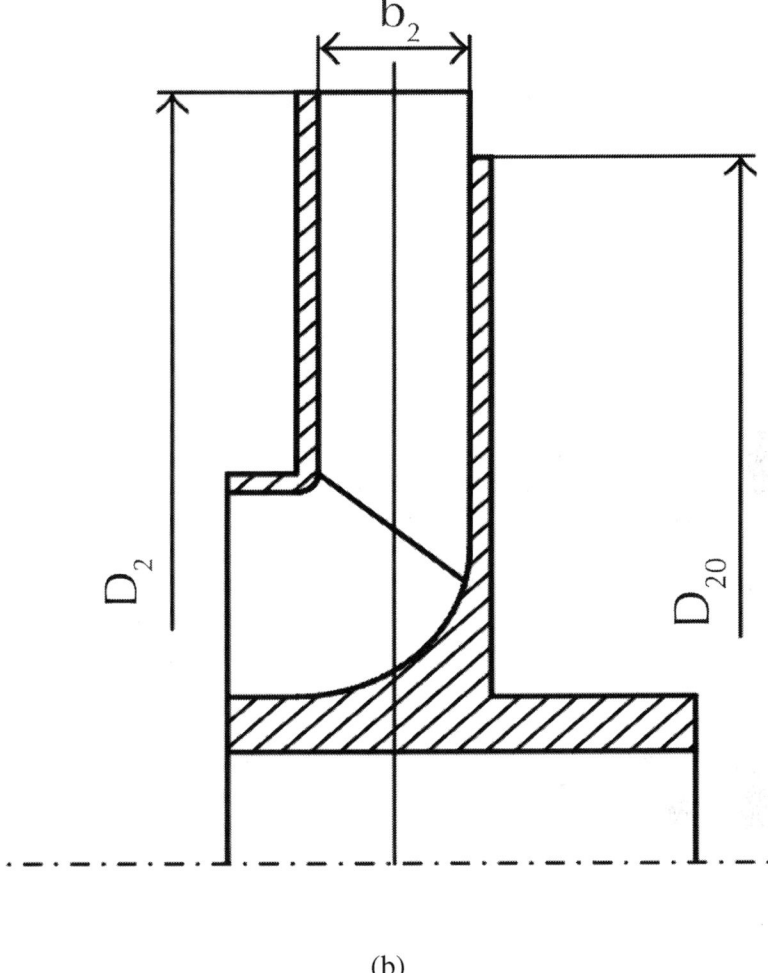

(b)

Figure 5: Two cutting proposals: (a) cutting blade outer edge and hub plate and (b) keeping blade and cutting hub plate.

Thus, for the first cutting proposal, the effective diameter of impeller is themean value of the shroud and the hub, $D_2 = (D_{21} + D_{20})/2$, as depicted in Figure 5(a). And the effective diameter for the second proposal is the shroud diameter, as shown in Figure 5(b). Then, the outer diameter of RGV is $D_4 \in [1.03, 1.08]\ D_2$. And the inlet meridional velocity at leading edge of RGV is

$$V_{m3} = \frac{4Q}{\pi\left(D_2{}^2 - D_{20}{}^2\right)} . \tag{1}$$

And the circumferential velocity at leading edge of RGV is

$$V_{u3} = \frac{gH/\eta_h}{u_2}, \tag{2}$$

where η_h is the impeller hydraulic efficiency, which is assumed to be known. Thus, with the meridional velocity V_{m3} and circumferential velocity V_{u3}, the RGV inlet flow angle at the streamline a is

$$\alpha_{3a} = \arctan\left(\frac{V_{m3}}{V_{u3}}\right). \tag{3}$$

With circumferential velocity decreasing, the inlet flow angle at streamline b and c is increased gradually. As the flow field information within RGV is insufficient, it is presumed that $\alpha_{3b} = \alpha_{3a}$ + const, and $\alpha_{3c} = \alpha_{3b}$ + const, where the const can be revised at design stage. Similarly, theRGVoutlet angle is presumed to be at $\alpha_4 \in [45°, 90°]$, the wrap angle is at $\phi \in [60°, 130°]$, and the RGV blade height $b_4 \in [1.05, 1.4]$ b_2. These parameters are flexible at design stage, which can be adjusted via the GUI of our in-house code NQSJ [41]. With these structural parameters, it is ready to draw the blade profile.

The profile drawing procedure is similar to the design of impeller blades [40]. Firstly, the blade angle distribution profile frominlet angle α_3 to outlet angle α_4 is constructed by aBezier curve asplotted inFigure 6(a), where a, b, c represent the blade angle distribution profile of three streamlines and d represents the dividing point. And the cubic Bezier curve is

$$\mathrm{Bez}\,(s) = P_0(1-s)^3 + 3P_1 s(1-s)^2$$
$$+ 3P_2 s^2(1-s) + P_3 s^3, \quad s \in [0,1], \tag{4}$$

where P_0 represents the inlet point, that is, points a, b, and c, and P_3 represents the outlet point, that is, point o, as shown in Figure 6(a). P_1 and P_2 are auxiliary points which are controlled by NQSJ. With the angle distribution profiles, the corresponding blade profiles are plotted point by point as shown in Figure 6(b).

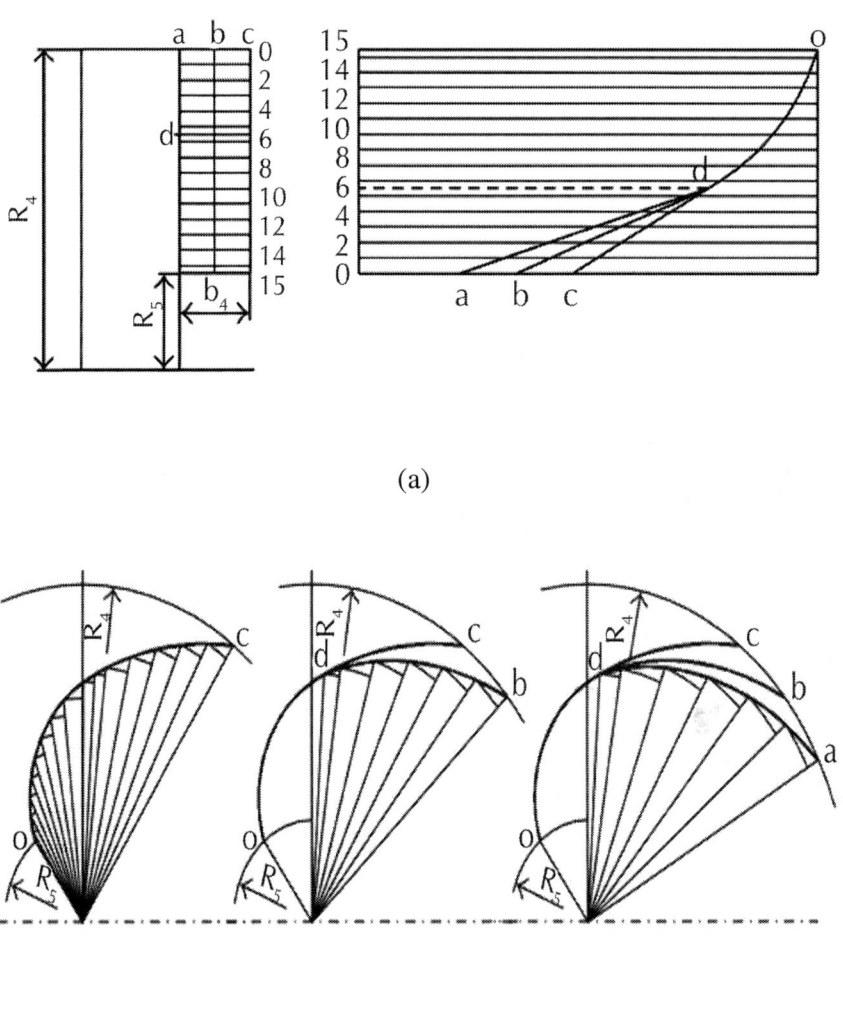

(a)

(b)

Figure 6: Blade profile drawing procedure: (a) blade angle distribution profiles and (b) blade profiles.

INSTANTANEOUS ROTOR-STATOR INTERACTION ANALYSIS

Pump Configurations

To examine the new RGV design method, a prototype pump 125QSJ10 is designed, which means the pump is used in a 125 mm (4 inch) deep well. The impeller discharge is cut using the first proposal. The major design parameters are listed in Tables 1 and 2.

Table 1: Design parameters of impeller

Q(m³/h)	H(m)	N (r/min)	Z_l	D_{21} (mm)	D_{20} (mm)	b_2 (mm)
10	8.5	2850	7	103	98	8

Table 2: Design parameters of RGV

Z_R	D_4 (mm)	B_4 (mm)	α_{3a} (°)	α_{3b} (°)	α_{3c} (°)	α_4 (°)	b_2 (°)
6	105	10	10	13.5	16.5	65	75

To shorten design cycle, a hydraulic design system NQSJ [40] is heavily used in the blade design and optimization procedures. And the main structure and such hydraulic components as impeller and RGV are depicted in Figure 7.

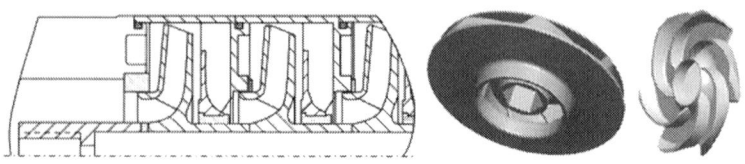

Figure 7: Schematic of 125QSJ10 and its impeller and RGV.

Numerical Configuration

Basic Equations

For incompressible flows, mass conservation equation is

$$\frac{\partial \rho}{\partial t} + \nabla \cdot (\rho \mathbf{u}) = 0. \tag{5}$$

For flows under noninertial rotating frame, momentum conservation equation is

$$\frac{\partial (\rho \mathbf{u})}{\partial t} + \nabla \cdot (\rho u \mathbf{u}) + \rho \left[2\boldsymbol{\omega} \times \mathbf{u} + \boldsymbol{\omega} \times \boldsymbol{\omega} \times \mathbf{r} \right]$$

$$= -\nabla p + \nabla \cdot \sigma, \tag{6}$$

where $\sigma = \mu(\partial V_i /\partial x_j + \partial V_j /\partial x_i)$, $2\omega \times u$ represents the Coriolis force and $\omega \times \omega \times r$ represents centrifugal force. And the previous equations applies for the multistage centrifugal pumps.

The RNG k-ε model is used for treatment of turbulence:

$$\frac{\partial k}{\partial t} + u_i \frac{\partial k}{\partial x_i} = \frac{\partial}{\partial x_j} \left[\alpha_k \frac{\mu_{\text{eff}}}{\rho} \frac{\partial k}{\partial x_j} \right]$$

$$+ \frac{\mu_t}{\rho} \left[\frac{\partial u_i}{\partial x_j} + \frac{\partial u_j}{\partial x_i} \right] \frac{\partial u_i}{\partial x_j} - \varepsilon,$$

$$\frac{\partial \varepsilon}{\partial t} + u_i \frac{\partial \varepsilon}{\partial x_i} = \frac{\partial}{\partial x_j} \left[\alpha_\varepsilon \frac{\mu_{\text{eff}}}{\rho} \frac{\partial \varepsilon}{\partial x_j} \right] \tag{7}$$

$$+ C_{1\varepsilon} \frac{\varepsilon}{k} \frac{\mu_t}{\rho} \left[\frac{\partial u_i}{\partial x_j} + \frac{\partial u_j}{\partial x_i} \right] \frac{\partial u_i}{\partial x_j} - C_{2\varepsilon}^* \frac{\varepsilon^2}{k},$$

$$C_{2\varepsilon}^* = C_{2\varepsilon} + \frac{C_\mu \eta^3 (1 - \eta/\eta_0)}{1 + \beta \eta^3},$$

where $\mu_{\text{eff}} = \mu + \mu_t$ represents effective viscosity, and $\mu_t = \rho C_\mu (k^2 /\varepsilon)$ represents turbulent viscosity. And $\eta = [2S_{ij} \cdot S_{ij}]^{1/2} (k/\varepsilon)$, where $S_{ij} = (1/2)[\partial u_i /\partial x_j + \partial u_j /\partial x_i]$ is fluid strain rate. And the model related constants are $C_\mu = 0.0845$, $\alpha_k = \alpha_\varepsilon = 1.393$, $C_{1\varepsilon} = 1.42$, $C_{2\varepsilon} = 1.68$, $\eta_0 = 4.38$, and $\beta = 0.012$. To date, a better choice of turbulence model for rotating flow within turbomachines is still not available. Relatively, the application of RNG k-ε model receives more support, as well as its accuracy.

CFD Setup

The calculation model is set up in the commercial package Ansys FLUENT to perform the simulation. A dual-stage calculation model is built to evaluate its performance. And there are four sets of interfaces between the stationary parts and the rotating parts, as shown in Figure 8.

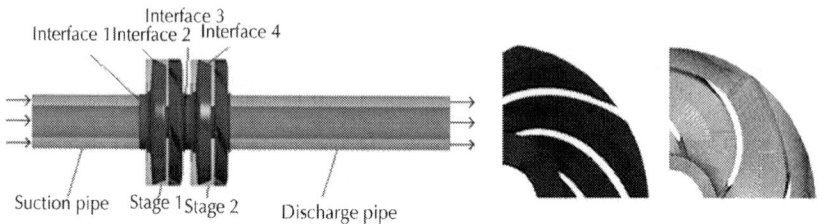

Figure 8: Dual-stage calculation model and numerical grids.

The mean velocity inlet boundary condition is prescribed on the suction, where the velocity is $U_{suc} = 4Q/\pi(D_1^2 - D_h^2)$, D_1 is the impeller suction diameter, and D_h is the impeller hub diameter. And the flow at discharge is supposed to be fully developed, so the Neumann boundary condition $\partial\varphi/\partial n = 0$ is imposed, φ represents such unknown variables as velocity, pressure, and turbulence scalars. No-slip boundary condition is imposed on the impeller and the guide vane wall boundaries, where $\varphi = 0$. And the turbulent flows are simulated by the RNG k-ε model and the wall function for near wall treatment.

Grid Irrelevance Analysis

The numerical mesh was generated by Ansys ICEM, where yplus $\in$ [30, 150]. Then, the steady-state simulation is performed to achieve such time-averaged characteristics as pump head, torque and efficiency. The stage head is

$$H = \frac{\left(P_{tot_{dis}} - P_{tot_{suc}}\right)}{\rho g},$$

(8)

where P_{totdis} represents the stage total pressure at discharge and P_{totsuc} represents the stage total pressure at suction. And the stage efficiency is

With steady-state simulations, the grid irrelevance is examined at three grid levels. And the obtained stage head and its estimated error are listed in Table 3.

Table 3

Number of cells	Stage head (m)	Error (%)
700,584	10.173	5.08
1,300,745	9.982	2.907
1,510,374	9.958	2.66

So in the present study, the magnitude of mesh size is around 1,500,000; thus, the grid irrelevance can be guaranteed.

Rotor-Stator Interaction

Instantaneous rotor-stator interaction rather than steady-state flow field reveals the real-time flow situation. Herein, the sliding mesh is used for the rotor-stator interaction analysis. And the pressure coefficient is used to evaluate its fluctuation as follows:

$$C_p = \frac{p - \bar{p}}{\rho u_2^2/2},$$

(10)

where $\bar{p}$ is the averaged pressure within a rotation and u_2 is the impeller wheel velocity.

As shown in Figure 9, pressure monitorings are conducted at ten different positions within the impeller and RGV.

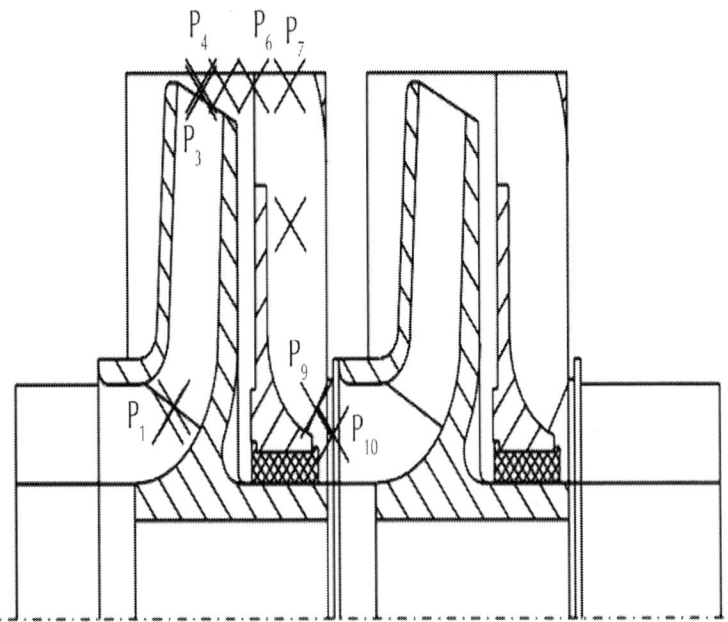

Figure 9: Pressure monitoring positions.

In Figures 10(a) and 10(b), with blade perturbation, the pressure fluctuation is formed within the impeller. It is cyclic symmetric at the leading edge. Inherently, the fluctuation runs seven peaks and valleys, equaling to the number of blades. And gradually, it is deviated from the cyclic symmetry by the intervention of RGV. With water outflowing from impeller, the fluctuation appears to be abnormal, as shown in Figure 10(c). But with the renormalization of RGV, it reappears cyclic, but only two peaks and valleys are retained as shown in Figure 10(d).

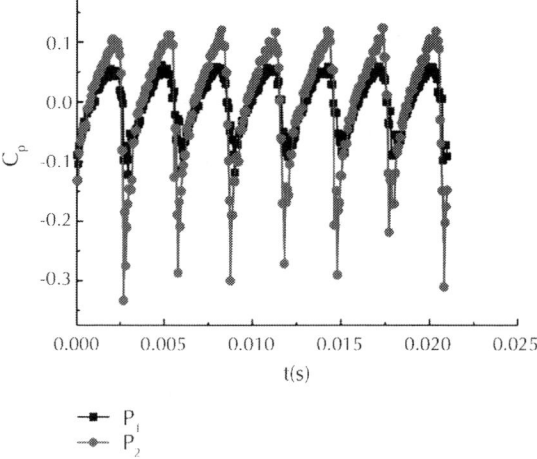

(a)

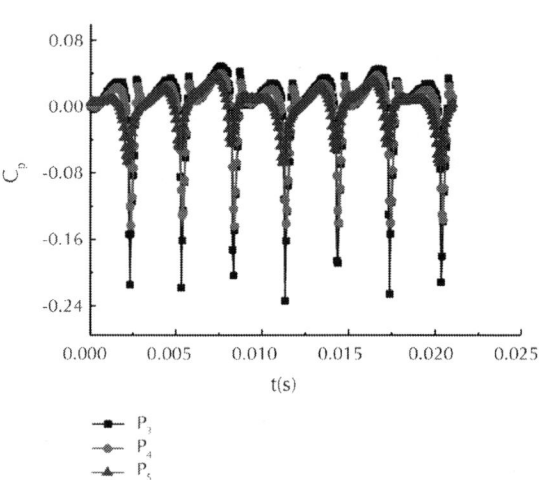

(b)

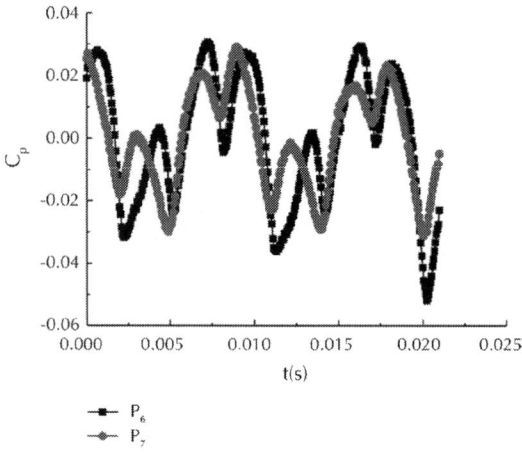

(c)

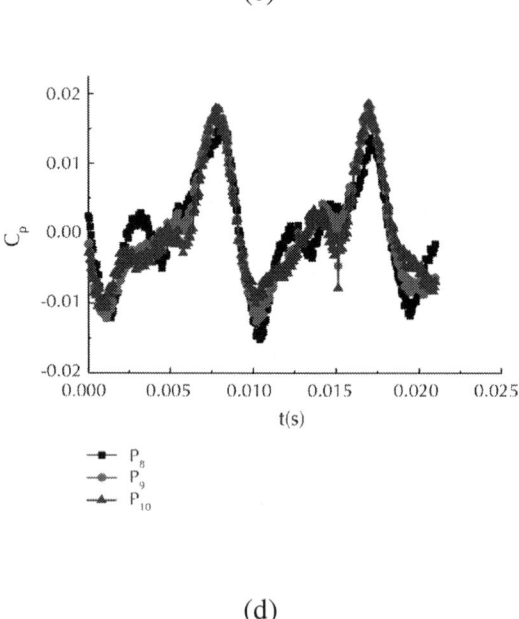

(d)

Figure 10: Pressure fluctuation at different locations: (a) impeller leading edge, (b) impeller trailing edge, (c) RGV leading edge, and (d) within RGV.

The frequency spectra are shown in Figure 11, which are obtained by FFT transformations. For multistage pumps, the impeller frequency is

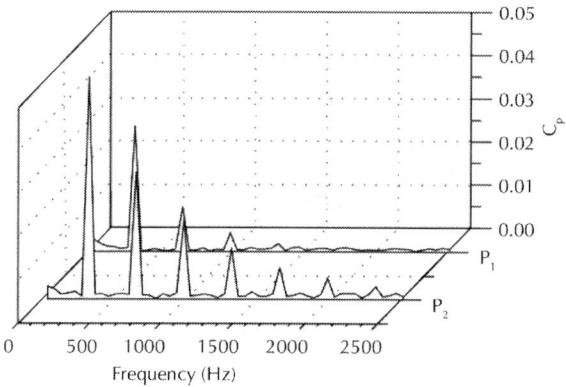

(a)

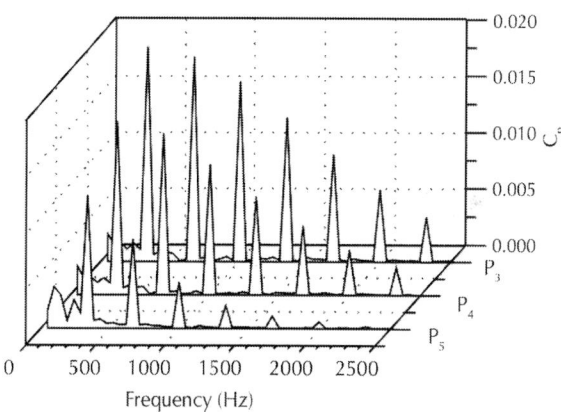

(b)

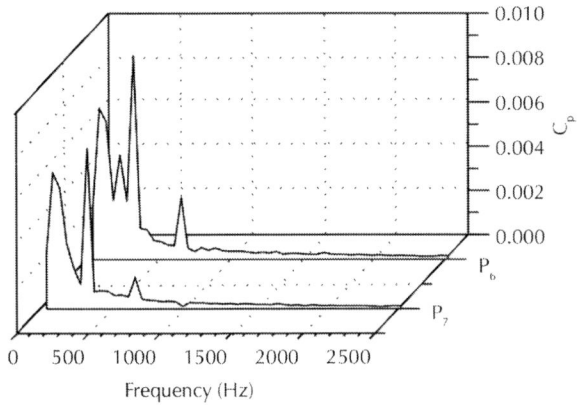

(c)

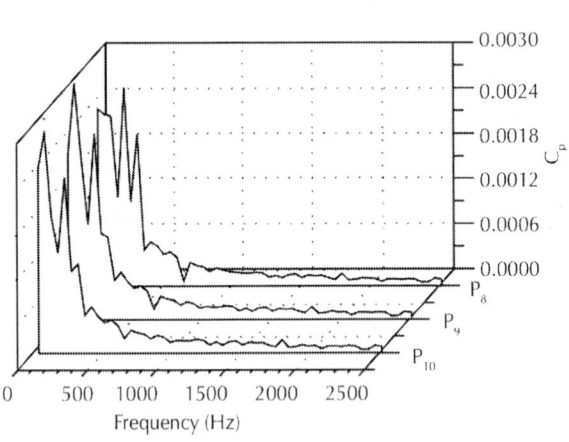

(d)

Figure 11: Frequency spectra at different locations: (a) impeller leading edge, (b) impeller trailing edge, (c) RGV leading edge, and (d) within RGV.

Substitute n = 2850 (r/min), Z_I = 7, into (10), f_I is 332.5Hz. In impeller, as above the fluctuation frequency is dominated by the impeller frequency, that is, 332.5Hz, as shown in Figures 11(a) and 11(b).With water outflowing from the impeller, the impeller frequency is even pronounced, but a secondary frequency appears with magnitude $n/60$, that is, 47.5Hz, which is the rotation frequency, as shown in Figure 11(c).Within RGV, the rotation frequency dominates, and the impeller frequency becomes the secondary frequency, as shown in Figure 11(d).

Prototype Pump Testing

To verify the previous design strategies, a four-stage pump is designed and tested. The testing is conducted at the pump test rig of the Fujian Academy of Mechanical Sciences. And the stage-averaged performance, such as the flow rate versus head and the flow rate versus efficiency curves are plotted in Figure 12.

Figure 12: Prototype pump and its characteristic curves.

In Figure 12, the results show that the peak efficiency is 57.5%, and the stage head is 9 m. And its efficiency is comparable to former multistage pumps, while commonly the stage head of former multistage pump is near 6.5 m. So the more head is needed, the greater cost-saving potential it will produce.

CONCLUSIONS

In this study, a compact structure is put forward and compared with the former structure of multistage centrifugal pumps. It is evident that its light weight and mobility will attract more prospective users. To meet this compact structure, a new RGV design is proposed. By using a triple-streamline surface shaping, a circumferential twisted RGV is developed.

To validate this design, the instantaneous rotor-stator interactions are investigated using the sliding mesh. And the pressure fluctuations are analyzed. The results show that the maximum fluctuation appears at the leading edge of RGV. And at the trailing edge of RGV the fluctuation is almost negligible. To check validity of this design, a prototype pump 125QSJ10 is developed and its characteristic curves are obtained. And its performance is comparable with the former multistage pump. But under a given downhole installation diameter, the compact structure with four stages can replace the former pump with six stages, clarifying its superiority.

On the other hand, in many areas where water resource is lacking, such as desert areas and arid areas, pumping water from underground is very inefficient with traditional reciprocating pumps, as well as in many oil-drilling applications; the new design would be an efficient alternative to reducing energy consumptions. Evidently, this compact multistage pump with new RGV is cost saving, which is attractive for the users.

REFERENCES

1.   A. Thakker, T. S. Dhanasekaran, and J. Ryan, "Experimental studies on effect of guide vane shape on performance of impulse turbine for wave energy conversion," Renewable Energy, vol. 30, no. 15, pp. 2203–2219, 2003.

2.   A. Thakker and T. S. Dhanasekaran, "Computed effect of guide vane shape on performance of impulse turbine for wave energy conversion," International Journal of Energy Research, vol. 29, no. 13, pp. 1245–1260, 2005.

3. L. M. C. Ferro, L. M. C. Gato, and A. F. O. Falcão, "Design and experimental validation of the inlet guide vane system of a mini hydraulic bulb-turbine," Renewable Energy, vol. 35, no. 9, pp. 1920–1928, 2010.

4. M. Govardhan and T. S. Dhanasekaran, "Effect of guide vanes on the performance of a self-rectifying air turbine with constant and variable chord rotors," Renewable Energy, vol. 26, no. 2, pp. 201–219, 2002.

5. Q. Li, H. Quan, R. Li, and D. Jiang, "Influences of guide vanes airfoil on hydraulic turbine runner performance," in Proceedings of the International Conference on Modern Hydraulic Engineering (CMHE ‹12), pp. 703–708, Nanjing, China, March 2012.

6. A. Mohseni, E. Goldhahn, R. A. Van den Braembussche, and J. R. Seume, "Novel IGV designs for centrifugal compressors and their interaction with the impeller," Journal of Turbomachinery, vol. 134, no. 2, Article ID 021006, 8 pages, 2012.

7. S. L. Gunter, S. A. Guillot, W. F. Ng, and S. T. Bailie, "A three-dimensional CFD design study of a circulation control inlet guide vane for a transonic compressor," in Proceedings of the 54th ASME Turbo Expo 2009, pp. 91–101, Orlando, Fla, USA, June 2009.

8. M. Hensges, "Simulation and optimization of an adjustable inlet guide vane for industrial turbo compressors," in Proceedings of the 53rd ASME Turbo Expo 2008, pp. 11–20, Berlin, Germany, June 2008.

9. M. Coppinger and E. Swain, "Performance prediction of an industrial centrifugal compressor inlet guide vane system," Proceedings of the Institution of Mechanical Engineers A, vol. 214, no. 2, pp. 153–164, 2000.

10. J. Fukutomi and R. Nakamura, "Performance and internal flow of cross-flow fan with inlet guide vane," JSME International Journal B, vol. 48, no. 4, pp. 763–769, 2005.

11. W. K. Chan, Y. W. Wong, S. C. M. Yu, and L. P. Chua, "A computational study of the effects of inlet guide vanes on the performance of a centrifugal blood pump," Artificial Organs, vol. 26, no. 6, pp. 534–542, 2002.

12. V. Chernoray, S. Ore, and J. Larsson, "Effect of geometry deviations on the aerodynamic performance of an outlet guide vane cascade," in Proceedings of the ASME Turbo Expo 2010: Power for Land, Sea, and Air, pp. 381–390, Glasgow, UK, June 2010.

13. T. Sonoda and H. Schreiber, "Aerodynamic characteristics of supercritical outlet guide vanes at low Reynolds number conditions," Journal of Turbomachinery, vol. 129, no. 4, pp. 694–704, 2007.

14. H. Posson, S. Moreau, and M. Roger, "Broadband noise prediction of fan outlet guide vane using a cascade response function," Journal of Sound and Vibration, vol. 330, no. 25, pp. 6153–6183, 2011.

15. C. Clemen, "Aero-mechanical optimisation of a structural fan outlet guide vane," Structural and Multidisciplinary Optimization, vol. 44, no. 1, pp. 125–136, 2011.

16. A. G. Barker and J. F. Carrotte, "Influence of compressor exit conditions on combustor annular diffusers, part 1: diffuser performance," Journal of Propulsion and Power, vol. 17, no. 3, pp. 678–685, 2001.

17. V. Cyrus and J. Polansky, "Numerical simulation of the flow pulsations origin in cascades of the rear blade rows in a gas turbine axial compressor using low calorific fuel," Journal of Turbomachinery, vol. 132, no. 3, Article ID 031012, 11 pages, 2010.

18. D. Kaya, "Experimental study on regaining the tangential velocity energy of axial flow pump," Energy Conversion and Management, vol. 44, no. 11, pp. 1817–1829, 2003.

19. W. Yan, D. Shi, Z. Luo, and Y. Lu, "Three-dimensional CFD study of liquid-solid flow behaviors in tubular loop polymerization reactors: the effect of guide vane," Chemical Engineering Science, vol. 66, no. 18, pp. 4127–4137, 2011.

20. A. Sulaiman and S. Gabin, "Flow through the return channel of a multistage centrifugal pump," in Proceedings of the 5th Conference on Fluid Machinery, vol. 2, pp. 1121–1132, Akademiai Kiado, Budapest, Hungary, 1975.

21. S. König and N. Petry, "Parker-type acoustic resonances in the return guide vane cascade of a centrifugal compressor—theoretical modeling and experimental verification," in Proceedings of the ASME Turbo Expo 2010: Power for Land, Sea, and Air, pp. 1687–1700, Glasgow, UK, June 2010.

22. M. Miyano, T. Kanemoto, D. Kawashima, A. Wada, T. Hara, and K. Sakoda, "Return vane installed in multistage centrifugal pump," The International Journal of Fluid Machinery and Systems, vol. 1, no. 1, pp. 57–63, 2008.

23. D. Kawashima, T. Kanemoto, K. Sakoda, et al., "Matching diffuser vane with return vane installed in multistage centrifugal pump," The International Journal of Fluid Machinery and Systems, vol. 1, no. 1, pp. 86–91, 2008.

24. J. D. Denton, "Some limitations of turbomachinery CFD," in Proceedings of the ASME Turbo Expo 2010: Power for Land, Sea, and Air, pp. 1–11, Glasgow, UK, June 2010.

25. T. Nagahara and Y. Inoue, "Investigation of hydraulic design for high performance multi-stage pump using CFD," in Proceedings of the ASME Fluids Engineering Division Summer Conference (FEDSM ‹09), pp. 417–424, Vail, Colo, USA, August 2009.

26. S. Huang, A. A. Mohamad, and K. Nandakumar, "Numerical simulation of unsteady flow in a multistage centrifugal pump using sliding mesh technique," Progress in Computational Fluid Dynamics, vol. 10, no. 4, pp. 239–245, 2010.

27. T. L. Rodgers and M. Cooke, "Rotor-stator devices: the role of shear and the stator," Chemical Engineering Research and Design, vol. 90, no. 3, pp. 323–327, 2012.

28. N. Sengil, "Performance increase in turbomolecular pumps with curved type blades," Vacuum, vol. 86, no. 11, pp. 1764–1769, 2012.

29. T. Chen and S. C. Li, "Numerical investigation of guide-plate induced pressure fluctuations on guide vanes of three gorges turbines," Journal of Fluids Engineering, vol. 133, no. 6, Article ID 061101, 10 pages, 2011.

30. J. Zhang, H. Zhu, C. Yang, Y. Li, and H. Wei, "Multi-objective shape optimization of helico-axial multiphase pump impeller based on NSGA-II and ANN," Energy Conversion and Management, vol. 52, no. 1, pp. 538–546, 2009.

31. D. Kaya, E. A. Yagmur, K. S. Yigit, F. C. Kilic, A. S. Eren, and C. Celik, "Energy efficiency in pumps,"Energy Conversion and Management, vol. 49, no. 6, pp. 1662–1673, 2008.

32. S. Sallem, M. Chaabene, and M. B. A. Kamoun, "Optimum energy management of a photovoltaic water pumping system," Energy Conversion and Management, vol. 50, no. 11, pp. 2728–2731, 2009.

33. M. Gölcü, Y. Pancar, and Y. Sekmen, "Energy saving in a deep well pump with splitter blade," Energy Conversion and Management, vol. 50, no. 11, pp. 2728–2731, 2009.

34. H. Roclawski and D.-H. Hellmann, "Rotor-stator-interaction of a radial centrifugal pump stage with minimum stage diameter," in Proceedings of the 4th WSEAS International Conference on Fluid Mechanics and Aerodynamics, pp. 301–308, Elounda, Greece, 2006.

35. W. Shi, Q. Zhang, and W. Lu, "Hydraulic design of new-type deep well pump and its flow calculation,"Journal of Jiangsu University, vol. 27, no. 6, pp. 528–531, 2006 (Chinese).

36. H. Roclawski and D.-H. Hellmann, "Numerical simulation of a radial multistage centrifugal pump," inProceedings of the 44th AIAA Aerospace Sciences Meeting and Exhibit, AIAA2006-1428, Elounda, Greece, January 2006.

37. H. Roclawski, A. Weiten, and D.-H. Hellmann, "Numerical investigation and optimization of a stator for a radial submersible pump stage with minimum stage diameter," in Proceedings of the ASME Joint U.S.-European Fluids Engineering Division Summer Meeting (FEDSM ‹06), FEDSM2006-98181, pp. 233–243, Miami, Fla, USA, July 2006.

38. W. Shi, W. Lu, H. Wang, and Q. Li, "Research on the theory and design methods of the new type submerible pump for deep well," in Proceedings of the ASME Fluids Engineering Division Summer Conference (FEDSM ‹09), pp. 91–97, Vail, Colo, USA, August 2009.

39. L. Zhou, W. D. Shi, W. G. Lu, et al., "Numerical investigations and performance experiments of a deep-well centrifugal pump with different diffusers," Journal of Fluids Engineering, vol. 134, no. 7, Article ID 071102, 8 pages, 2012.

40. J. F. Gulich, Centrifugal Pumps, Springer, New York, NY, USA, 2007.

41. Q. H. Zhang, Y. Xu, W. D. Shi, et al., "Research and development on the hydraulic design system of the guide vanes of multistage centrifugal pumps," Applied Mechanics and Materials, vol. 197, pp. 24–30, 2012.

Chapter 6

Research on Three-Dimensional Unsteady Turbulent Flow in Multistage Centrifugal Pump and Performance Prediction Based on CFD

Zhi-jian Wang[1], Jian-she Zheng[1], Lu-lu Li[2], and Shuai Luo[1]

[1]School of Mechatronics Engineering, Shenyang Aerospace University, Shenyang, Liaoning 110136, China

[2]Haicheng Suprasuny Pump Co., Ltd., Haicheng, Liaoning 114216, China

ABSTRACT

The three-dimensional flow physical model of any stage of the 20BZ4 multistage centrifugal pump is built which includes inlet region, impeller flow region, guide-vane flow region and exit region. The three-dimensional unsteady turbulent flow numerical model is created based on Navier-Stoke solver and standard k-ε turbulent equations. The method of multi reference frame (MRF) and SIMPLE algorithm are used to simulate the flow in multistage centrifugal pump based on FLUENT software. The distributions of relative velocity, absolute velocity, static pressure, and total pressure in guide vanes and impellers under design condition are analyzed. The simulation results show that the flow in impeller is mostly uniform, without eddy, backflow, and separation flow, and jet-wake phenomenon appears only along individual blades. There is secondary flow at blade end and exit of guide vane. Due to the different blade numbers of guide vane and impeller, the total pressure distribution is asymmetric. This paper also simulates the flow under different working conditions to predict the hydraulic performances of centrifugal pump and external characteristics including flow-lift, flow-shaft power, and flow-efficiency are attained. The simulation results are compared with the experimental results, and because of the mechanical losses and volume loss ignored, there is a little difference between them.

INTRODUCTION

Pumps are widely used in many fields and the average electric power consumption is about 20.9% of the total consumption every year in China [1]. Because of the low level of manufacture and design of pumps, the efficiency of domestic pumps is about 10% lower than that of the developed countries. Among the pumps, the centrifugal ones are most widely applied, but there are many problems such as low efficiency, operated under off-design conditions, and low cavitations performance. Therefore, it will have very important practical significance to study the internal flow of centrifugal pumps in order to optimize the structure of main parts, improve the hydraulic performance, increase the efficiency and avoid being operated under off-design conditions, and thus reach the goal of increasing efficiency and saving energy.

Due to the complex shape of flow channel, high-speed-rotating viscous fluid and the interaction between moving and stationary parts, the flow in centrifugal pumps is a three-dimensional, viscid, and unsteady complex flow. It becomes more and more popular to investigate the internal flow of the centrifugal pump based on computational fluid dynamic (CFD) owing to the short design time, low price, being observed directly, and making up the deficiency of traditional design methods. With the rapid development of the computer technology, CFD has been one of the main methods to study the flow in the centrifugal pump. Subsequently, it will be possible to design high-efficiency and energy-saving pumps and create huge social and economic benefits. Si and Dike [2] simulated the whole flow field of sectional multistage pump and the simulation was performed in a multiple reference frame and standard k-ε turbulence model. Li et al. [3] combined sliding-mesh and moving-mesh methods to simulate internal flow during starting procedure of the single-stage pump. Liu and Wang [4] carried out computer-aided analysis on internal flow of stamping and welding centrifugal pump impeller based on CFD using ANSYS CFX and explored the flow mechanism in impeller. Barrio et al. [5] simulated internal flow of centrifugal pump through CFD such that they could predict radical force and torsion suffered by impeller. Jafarzadeh et al. [6] simulated fluid flow of low-specific-speed ratio centrifugal pump. Asuaje et al. [7] performed a 3D-CFD simulation of impeller and volute of a centrifugal pump using CFX code with a specific speed of 32 and found velocity and pressure fields for different flow rates and radial thrust on the pump shaft. Cui et al. [8] investigated the effect of number of splitting blades for long, mid, and short blades using a one-equation turbulent model. Their results show that the bulk flow in the impeller has an important influence on the pump performance. Anagnostopoulos [9] simulated 3D turbulent flow in a radial pump impeller for a constant rotational speed of 1500 rpm based on the solution of the RANS equations. Few of the previous works involved study of 3D modeling within a full domain considering interaction between rotor and stator of a high-speed multistage centrifugal pump using various turbulence models.

This paper uses commercial CFD software FLUENT, standard k-ε turbulent model and multiply reference frame to perform numerical modeling of the full three-dimensional fluid field for any stage of 20BZ4 multistage centrifugal pump which includes the flow region of import channel, impellers, guide vanes, and exit channel. The pressure and velocity distributions in the pump under design condition are obtained and the numerical performance curves are compared with the experimental ones. It will provide theoretical basis for further optimizing the structures and improving the performances of centrifugal pump.

NUMERICAL SIMULATION AND METHOD

Physical Model

Figure 1 show the sketch map and flow route in any stage unit of 20BZ4 multi-stage centrifugal pump, which includes guide plates, impellers, and guide vanes. The impeller is made up of front end plate, back end plate and blades. The blades are equipped between front and back end plates, and the number of blades is 5. The structure of guide vanes is radial and the number of blades is 7. It is made up of positive guide vane and negative guide vane. Positive vane can collect fluid and transform kinetic energy into pressure energy, while negative vane can change flow direction and transmit the fluid into next unit with the required speed and circulation. Guide plate can reduce reflux effectively and make uniform and stable flow velocity of the fluid into the impeller. Fluid flows downward through guide plate, then through the flow runner of impeller into guide vane and finally goes into the next pump unit from guide plate.

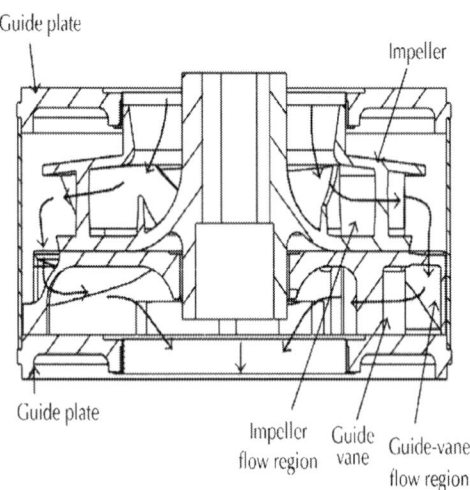

Figure 1: Sketch of centrifugal pump.

The impeller inlet and guide-vane outlet are extended, respectively, in order to ensure stable convergence of internal flow field. The physical model includes inlet region, impeller flow region, guide-vane flow region, and exit region. Figure 2 shows the flow region model of impeller and guide-vane. Structured grids are used in inlet region and exit region because of the cylindrical shape and the numbers of grids are 71607 and 69564, respectively. Unstructured grids are used to mesh impeller and guide-vane flow regions and the numbers of grids are 187561 and 133108, respectively Figures 3 and 4 show the grids of impeller flow region and guide-vane flow region, respectively.

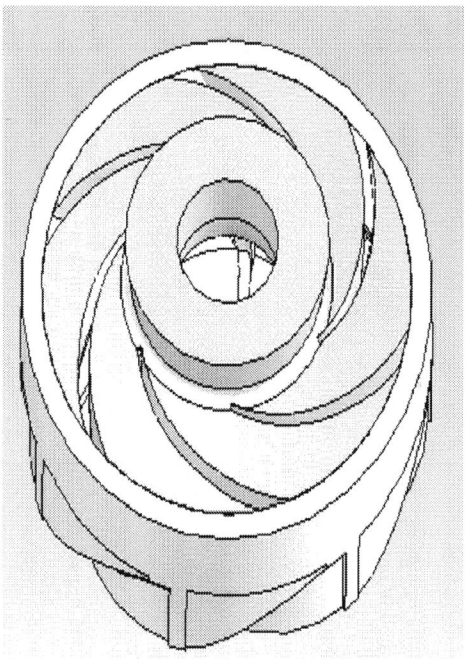

Figure 2: Flow region of impeller and guide vane.

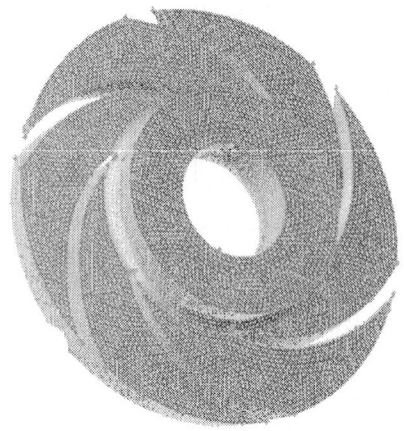

Figure 3: Grids of impeller flow region.

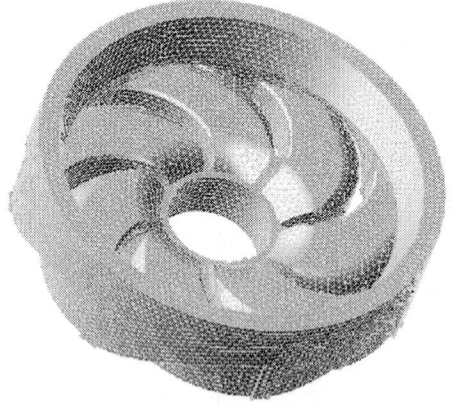

Figure 4: Grids of guide-vane flow region.

Governing Equations and Turbulence Model

The internal flow of centrifugal pump is a three-dimensional, viscous, and unsteady turbulent flow and flow law follows Navier-Stokes equation. Because the heat exchange is very little in centrifugal pump, energy conservation equation is not considered, and only mass conservation equation and momentum conservation equation need to be solved.

Mass conservation equation is as follows:

$$\frac{\partial \rho}{\partial t} + \frac{\partial}{\partial x_i}\left(\rho u_i\right) = 0.$$

(1)

Momentum conservation equation is as follows:

$$\frac{\partial}{\partial t}\left(\rho u_i\right) + \frac{\partial}{\partial x_j}\left(\rho u_i u_j\right) = -\frac{\partial p}{\partial x_i} + \frac{\partial}{\partial x_j}\left[\mu\frac{\partial u_i}{\partial x_j} - \overline{\rho u_i' u_j'}\right] + S_i,$$

(2)

Where ρ is fluid density, u is velocity, p is pressure, t is time, μ is dynamic viscosity, S is source item, and $\overline{\rho u_i' u_j'}$ is the Reynolds stress. x_i and x_j are the coordinates of x, y, and z, and $x_i \neq x_j$.

Standard k-ε turbulence model is used. Turbulence kinetic energy k equation is as follows:

$$\frac{\partial(\rho k)}{\partial t} + \frac{\partial(\rho k u_i)}{\partial x_i} = \frac{\partial}{\partial x_j}\left[\left(\mu + \frac{\mu_t}{\sigma_k}\right)\frac{\partial k}{\partial x_j}\right] + G_k - \rho\varepsilon.$$

(3)

Dissipation rate ε equation is as follows:

$$\frac{\partial(\rho\varepsilon)}{\partial t} + \frac{\partial(\rho\varepsilon u_i)}{\partial x_i} = \frac{\partial}{\partial x_j}\left[\left(\mu + \frac{\mu_t}{\sigma_s}\right)\frac{\partial\varepsilon}{\partial x_j}\right]$$

$$+ \frac{C_{1\varepsilon}\varepsilon}{k}G_k - C_{2\varepsilon}\rho\frac{\varepsilon^2}{k},$$

(4)

Where G_k is production term of turbulence energy k produced by average velocity gradient, $c_{1\varepsilon}$, $c_{2\varepsilon}$, and $c_{3\varepsilon}$ are empirical constants, σ_k and σ_s are Prandtl numbers of turbulence kinetic energy k and dissipation rate ε, and turbulence viscosity is defined as

$$\mu_t = \rho C_\mu\frac{k^2}{\varepsilon},$$

(5)

Where C_μ is the empirical constant.

Boundary Conditions and Numerical Model

Inlet Boundary Conditions

Velocity inlet surface, where velocity and other scalars are defined, is chosen as the inlet boundary. Inlet velocity can be calculated by

$$u_{in} = \frac{Q}{\rho \pi \left(r_1^2 - r_2^2 \right)},$$

(6)

Where Q is flow, and r_1 and r_2 are inlet cross section radii. Inlet turbulence energy k is calculated as

$$k_{in} = 0.005 u_{in}^2.$$

(7)

Inlet dissipation rate ε is calculated as

$$\varepsilon_{in} = \frac{C_\mu k_{in}^{3/2}}{l_{in}},$$

(8)

Where l_{in} is inlet mixing length, D is inlet equivalent diameter, and $l_{in}=0.5D$.

Outlet Boundary Conditions

The exit is set as outflow boundary which is mainly used where the exit flow is under full-developed state. The outlet velocity u_{out}, turbulence kinetic energy k_{out}, and dissipation rate ε_{out} are described in the following equations:

$$\frac{\partial u_{i(\text{out})}}{\partial n} = 0 \quad (i = 1, 2, 3, \ldots),$$

$$\frac{\partial k_{\text{out}}}{\partial n} = 0,$$

$$\frac{\partial \varepsilon_{\text{out}}}{\partial n} = 0, \tag{9}$$

Where n is the unit vector orthogonal to exit boundary.

Wall Boundary Conditions

No-slipping wall boundary conditions are assumed on the wall. The impeller boundary, front and back end plates are set as rotating wall, and other walls are stationary. Because the Reynolds number near the walls is small and standard k-ε model is not appropriate to turbulent boundary layer region, logarithmic wall function is used.

Numerical Method

Multiple reference frame (MRF) is used in FLUENT and unsteady problem can be transferred into steady problem. Steady calculation is done in stator region, while centrifugal force and Coriolis force are calculated in rotor region in inertial frame and inner grids keep stationary during calculation. Flow parameters are switched between the interfaces of impellers and guide vanes in order to keep continuity of interfaces.

SIMPLE algorithm is used to couple pressure with velocity, and segregated solver and standard discrete scheme are chosen. First order upwind scheme is used to solve momentum conservation equation, turbulence energy equation, and dissipation rate equation. Underrelaxation factor controls the convergence speed and is properly updated based on actual convergence condition.

Design Condition of Pump and Fluid Properties

The design condition of pump and fluid physical properties are shown in Table 1.

Table 1: Design condition and fluid physical properties

Flow (m³/h)	Rotational speed (r/min)	Atmospheric pressure (Pa)	Medium density (kg/m³)	Dynamic viscosity (Pa·s)
20	2850	101325	998.2	1.003×10^{-3}

NUMERICAL RESULTS

Velocity Distribution

Figure 5 shows the relative velocity vector of x=0 section in the middle of front end plate and back end plate of impellers under design condition. It can be seen from the figure that the flow is uniform in most fluid region of impellers without eddy, backflow, and separation flow. Jet-wake phenomenon happens only along individual blades. The flow velocity increases gradually from the inlet of impeller to exit, being slowest at the inlet and fastest at the exit. Because of diffusion function of guide blades, kinetic energy of high-speed fluid is transferred into pressure energy, and velocity becomes lower when fluid goes into the guide blade. Meanwhile a part of the energy is lost when the high-speed fluid flowing out of impellers collides into the pump case. The figure also shows that the relative velocity of suction surface is lower than that of pressure surface on the same radius surface. The pressure difference which is produced on the two sides of impellers due to the asymmetry creates the moment of resistance which is overcome by the prime mover to work on the spindle.

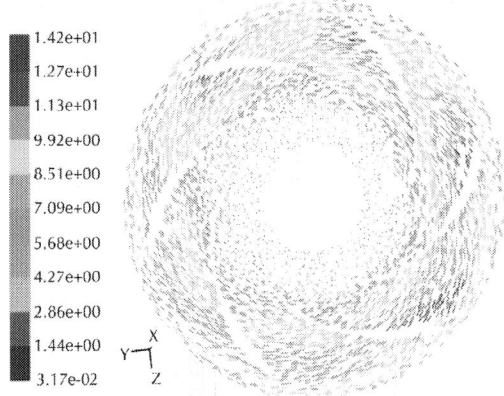

Figure 5: Relative velocity vector of impellers.

Figure 6 shows the absolute velocity of guide vanes under design condition. From the figure we can see that the velocity is fastest at the inlet of guide vane and slowest at the exit. The guide vanes transfer the kinetic energy of fluid into pressure energy. As a result, the velocity decreases gradually along the direction of the flow in the guide vane and secondary flow appears at blade end and exit.

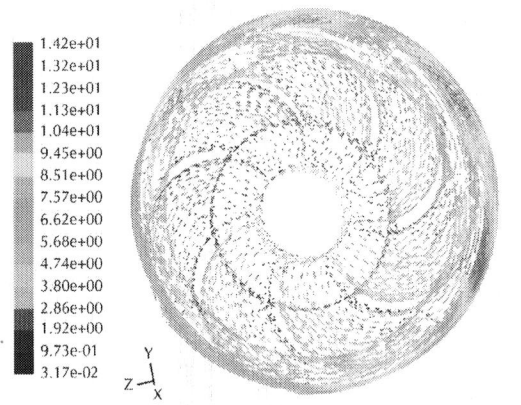

Figure 6: Absolute velocity vector of guide vanes.

Static Pressure Distribution

Figure 7 shows the static pressure distribution on the centrifugal pump impellers of x=0 section. It shows that the static pressure increases gradually and is ladder-like uniform distribution. The minimum pressure area appears at the suction surface of impeller inlet. The fluid can get the kinetic energy driven by impellers when it enters into the impeller flow channel vertically, but because the velocity direction changes quickly and some energy gets lost when the fluid collides into the impeller front end, cavitations could happen in these low-pressure areas. Figure 8 shows the static pressure distribution on suction surface and pressure surface of impellers, respectively. The pressure which shows ladder-like distribution gradually increases along the flow direction on both pressure and suction surfaces. The pressure on pressure surface is higher than that of suction surface and the pressure difference causes the moment of resistance on rotating axis. At the inlet of suction surface the pressure is lowest and cavitations may happen here.

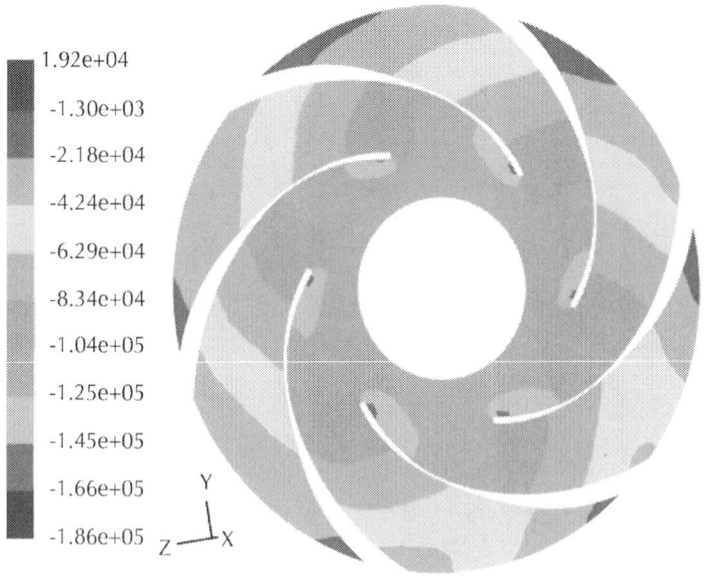

Figure 7: Static pressure distribution on impellers.

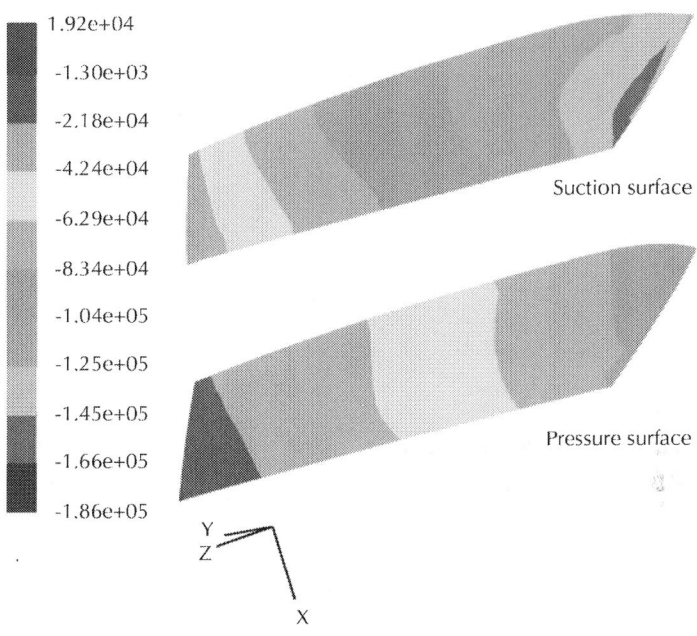

1.92e+04
-1.30e+03
-2.18e+04
-4.24e+04
-6.29e+04
-8.34e+04
-1.04e+05
-1.25e+05
-1.45e+05
-1.66e+05
-1.86e+05

Suction surface

Pressure surface

Y
Z
X

Figure 8: Static pressure distribution on suction surface and pressure surface of impellers.

Figure 9 shows the static pressure distribution on guide vanes of centrifugal pump under the design condition. It shows that the pressure increases gradually along the flow direction and reaches the maximum value at the exit of guide vanes. The function of guide vanes is to collect the high-speed fluid and then transfers kinetic energy of the fluid into pressure energy. Because of the crash between high-speed fluid from impellers and pump case, local low pressure appears in the interface of impellers and guide vanes, but it disappears when the fluid enters into the guide vanes. Figure 10 shows the total pressure distribution of the centrifugal pump on x=0 section. It shows that the total pressure increases gradually which is ladder-like uniform distribution when fluid flows from the impeller inlet exit, and then enters into the guide blade. It displays different pressure distribution at the impeller export and the entrance of guide vanes because the blade number of the guide vane is 7 and that of the impellers is 6. Due to the different blade number, the relative location of different flow channels displays asymmetric distribution when the impellers rotate.

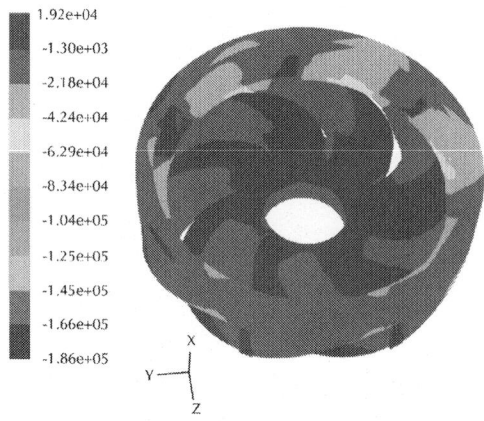

Figure 9: Static pressure distribution on guide vanes.

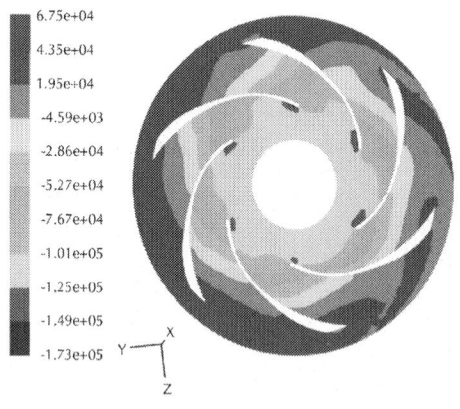

Figure 10: Total pressure distribution of centrifugal pump.

PERFORMANCE PREDICTION BASED ON NUMERICAL SIMULATION

In order to predict the hydraulic performances of centrifugal pump, the external characteristics including flow, shaft power, lift, and efficiency are calculated.

The flow of inlet surface Q in the centrifugal pump is defined as follows:

$$Q = \int_A (\rho \vec{v} \cdot \hat{n}) \, dA,$$

(10)

Where A is the area of the inlet or exit of the centrifugal pump $\vec{v}$ is the velocity vector of the calculation element, ρ is fluid density, and $\hat{n}$ is direction vector on the inlet surface or the exit surface.

The total pressure on the inlet and exit surfaces is respectively defined by the pattern of the mass average value as follows:

$$P_i = \frac{\int_A (\rho p_t \, |\vec{v} \cdot \hat{n}|) \, dA}{\int_A (\rho \, |\vec{v} \cdot \hat{n}|) \, dA},$$

(11)

Where p_t is the total pressure of the calculation element.

The lift of centrifugal pump is shown as follows:

$$H = \frac{P_{out} - P_{in}}{\rho g} + \frac{v_{out}^2 - v_{in}^2}{2g} + \Delta Z,$$

(12)

Where P_{in} and P_{out} are, respectively, the total pressure of the inlet and exit ΔZ is the vertical distance between the inlet and exit. v_{in} and v_{out} are, respectively, the speed of the inlet and exit. g is gravity acceleration.

The shaft power is calculated as follows:

$$P = Mw,$$

$$w = \frac{2\pi n}{60},$$

(13)

Where M is the total moment of pressure surface, suction surface, and front and back end plates around z axis, n is the rotated speed, and w is the angular velocity.

The centrifugal pump efficiency is shown as follows:

$$\eta = \frac{\rho g Q H}{Mw}.$$

(14)

In addition to the design condition, the paper simulates different flow conditions of 0.5Q, 0.7Q, 0.8Q, 0.9Q, 1.1Q, 1.2Q, 1.3Q, and 1.5Q to attain the lift, shaft power, and efficiency, as shown in Table 2.

Table 2: Performance data under off-design conditions of simulation

Flow (m³/h)	0.5Q	0.7Q	0.8Q	0.9Q	Q	1.1Q	1.2Q	1.3Q	1.5Q
Lift (m)	10.57	9.89	9.57	9.34	8.73	8.05	7.48	6.38	5.29
Shaft power (w)	546.3	617.9	634.7	666.5	686.4	711.7	763.3	760.7	856
Efficiency (%)	52.7	61.2	65.9	68.7	69.3	67.7	64	59.4	50.5

EXPERIMENT VERIFICATION

In order to verify the reliability of the results of numerical simulation, experiments are designed to test the flow, lift, shaft power, and efficiency of the 20BZ4 centrifugal pump. Figures 11, 12, and 13 shows, respectively, the experimental characteristic curves of flow and lift, flow and shaft power, and flow and efficiency The figures show that there are certain differences between the experimental results and numerical results. When the pump physical model is built, the gap region between front and back plates and case is ignored, so the rotation of pump is accompanied by a volume loss. Furthermore mechanical loss such as bearing friction loss and disc loss are also ignored.

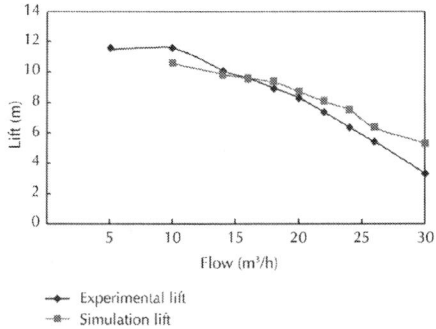

Figure 11: Performance curves of flow-lift.

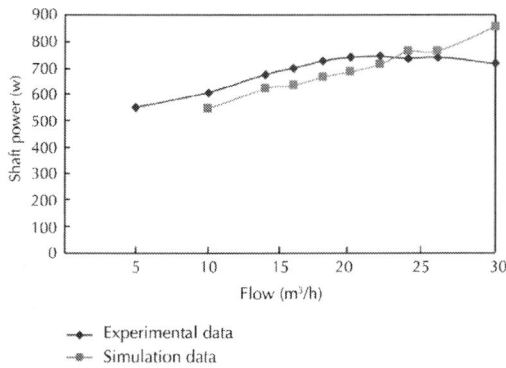

Figure 12: Performance curves of flow-shaft power.

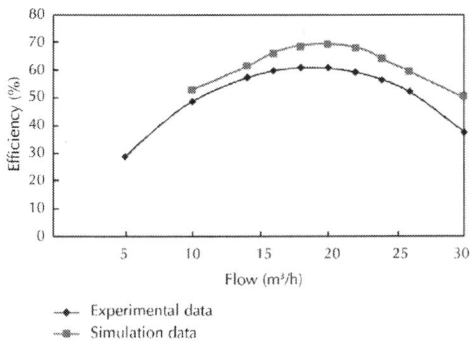

Figure 13: Performance curves of flow-efficiency.

Figure 11 shows the relation curve between the flow and lift. The simulation curve has no hump, and it demonstrates that the centrifugal pump has good performances. The trends of experimental curve and simulation curve are consistent. But in addition to the design condition, the experimental data and calculated data in the high flow and low flow have larger difference. MRF is a kind of assumed steady calculation flow model relative to the design condition, so the unsteady factors of flow field are fewer near the design condition and the calculated data and experimental data are consistent. However under off-design conditions, there are many unsteady factors such as the crash between the fluid and pump shell and blade boundary layer separation, which results in difference between calculated data and experimental data.

Figure 12 shows the relation curve between the flow and shaft power. Because the calculated moment is lower than the experimental moment, so the shaft power of numerical simulation is lower than that of experiment. However under high flow conditions, the shaft power of numerical simulation is higher than that of experiment, which is because the relative ideal numerical model is used and the distribution of the unsteady factors in the flow is not taken into account.

Figure 13 shows the relation curve between the flow and efficiency. It demonstrates that the curve first goes up and then down and it becomes relatively flat near the region of design condition. The flow region of high efficiency is wide which demonstrates that the performance is stable around the design condition. When the pump physical model is built, the gap region between front and back plates and case is ignored, so the rotation of pump is accompanied by a volume loss. Furthermore mechanical loss such as bearing friction loss and disc loss are also ignored. The actual losses cause the efficiency of numerical simulation to be higher than that of experiment, which can be seen from the figure.

CONCLUSIONS

- Complicated three-dimensional flow model is built including inlet region, impeller flow region, guide-vane flow region, and exit region to simulate flow in 20BZ4 multi-stage centrifugal pump. The method of multi reference frame (MRF) is used to model rotating blades and stationary blades by FLUENT.

- The simulation results show that the flow in impellers is mostly uniform, no eddy, backflow, and separation flow. The Jet-wake along some blades influences the efficiency. There is secondary flow at blade end and exit of guide vanes. The pressure on pressure surface is higher than that of suction surface and the pressure difference causes the moment of resistance on rotating axis. At the inlet of suction surface the pressure is lowest and cavitations may happen there.

- Besides design condition, six off-design conditions are set to predict the external characteristics of hydraulic performances. The comparison between experimental data and simulation data shows that the experimental curve agrees well with the simulation curve under design condition, but under off-design conditions the unsteady factors of flow field influence the precision. The actual losses cause the efficiency of numerical simulation to be higher than that of experiment.

REFERENCES

1. J. Shi, "The energy conservation improvement and prospect of the centrifugal pump," General Machinery, vol. 9, pp. 24–28, 2012.

2. H. Si and S. Dike, "Numerical simulation of the three-dimensional flow field in a multistage centrifugal pump and its performance prediction," Mechanical Science and Technology, vol. 29, no. 6, pp. 706–708, 2010.

3. Z. Li, D. Wu, L. Wang, and B. Huang, "Numerical simulation on internal flow of centrifugal pump during transient operation," Journal of Engineering Thermophysics, vol. 30, no. 5, pp. 781–783, 2009.

4. Y. Liu and G. Wang, "Computer-aided analysis on inner flow in stamping and welding multistage centrifugal pump's impellers," Chinese Journal of Mechanical Engineering, vol. 43, no. 8, pp. 207–211, 2007.

5. R. Barrio, J. Fernndez, E. Blanco, and J. Parrondo, "Estimation of radial load in centrifugal pumps using computational fluid dynamics," European Journal of Mechanics B, vol. 30, no. 3, pp. 316–324, 2011.

6.   B. Jafarzadeh, A. Hajari, M. M. Alishahi, and M. H. Akbari, "The flow simulation of a low-specific-speed high-speed centrifugal pump," Applied Mathematical Modelling, vol. 35, no. 1, pp. 242–249, 2011.·

7.   M. Asuaje, F. Bakir, S. Kouidri, F. Kenyery, and R. Rey, "Numerical modelization of the flow in centrifugal pump: volute influence in velocity and pressure fields," International Journal of Rotating Machinery, vol. 2005, no. 3, pp. 244–255, 2005.

8.   B. Cui, Z. Zhu, J. Zhang, and Y. Chen, "The flow simulation and ex perimental study of low-specific-speed high-speed complex centrifugal impellers," Chinese Journal of Chemical Engineering, vol. 14, no. 4, pp. 435–441, 2006.

9.   J. S. Anagnostopoulos, "Numerical calculation of the flow in a centrifugal pump impeller using Cartesian grid," in Proceedings of the 2nd WSEAS International Conference on Applied and Theoretical Mechanics, pp. 20–22, Venice, Italy, 2006.

Chapter 7

Optimization and Analysis of Centrifugal Pump considering Fluid-Structure Interaction

Yu Zhang[1], Sanbao Hu[2], Yunqing Zhang[3], and
Liping Chen[3]

[1]Wuhan Second Ship Design and Research Institute, Wuhan, Hubei 430064, China

[2]Hubei Key Laboratory of Advanced Technology of Automobile Parts, Wuhan University of Technology, Wuhan, Hubei 430074, China

[3]Center for Computer-Aided Design, School of Mechanical Science & Engineering, Huazhong University of Science & Technology, Wuhan, Hubei 430074, China

ABSTRACT

This paper presents the optimization of vibrations of centrifugal pump considering fluid-structure interaction (FSI). A set of centrifugal pumps with various blade shapes were studied using FSI method, in order to investigate the transient vibration performance. The Kriging model, based on the results of the FSI simulations, was established to approximate the relationship between the geometrical parameters of pump impeller and the root mean square (RMS) values of the displacement response at the pump bearing block. Hence, multi-island genetic algorithm (MIGA) has been implemented to minimize the RMS value of the impeller displacement. A prototype of centrifugal pump has been manufactured and an experimental validation of the optimization results has been carried out. The comparison among results of Kriging surrogate model, FSI simulation, and experimental test showed a good consistency of the three approaches. Finally, the transient mechanical behavior of pump impeller has been investigated using FSI method based on the optimized geometry parameters of pump impeller.

INTRODUCTION

Centrifugal pumps provide the energy to move fluids through piping systems, including equipment, piping, and fittings and through elevation changes in open systems. Centrifugal pumps have been widely used in various industrial applications, such as oil and gas, agriculture, chemistry, and marine industry as well as metallurgy. Because of the customers' increasing demands of high-quality pump, optimization design of centrifugal pump plays an important role in pump industry, and there have been many efforts to optimize the performance of centrifugal pump in recent years. Anagnostopoulos [1] proposed an optimization algorithm based on unconstrained gradient method to find the impeller geometry that could maximize the pump efficiency among a set of blade angles. Zhou et al. [2] optimized the geometric shape of the centrifugal impeller using orthogonal experiment method to improve the performance of the centrifugal pump. Derakhshan et al. [3] presented the incomplete sensitivities approach and genetic algorithms to obtain a higher efficiency by redesigning the shape of impeller blades. Papierski and Blaszczyk [4] decomposed the optimization design of centrifugal pump into two levels to maximize the efficiency and simultaneously minimize required net positive suction head (NPSHr). These researches mainly focus on optimizing the performance data of centrifugal pump, such as head, efficiency, or NPSHr. However, the vibration performance is important especially for high-pressure centrifugal pump.

The vibration that occurs while centrifugal pump works can cause fatigue and damage of pump components and weaken the operation stability. Vibrations of centrifugal pump have attracted interest of researchers. For example Hodkiewicz and Norton [5] investigated the influence of different flow rates on the vibration performance of double-suction centrifugal pump. Guo and Maruta [6] presented an experimental study of the pressure fluctuation and the impeller vibration in a centrifugal pump with some vaned diffusers. Rodriguez et al. [7] developed a theoretical analysis approach to investigate the vibrating frequencies in the vibration of centrifugal pump induced by the rotor stator interaction (RSI). Wang et al. [8] studied the structural dynamics characteristics and vibrations of pump volute casing for a double-suction centrifugal pump using a fluid-structure coupling interface model.

Studies agree in considering the fluid-structure interaction (FSI) as the source of the highest vibration levels in large centrifugal pumps. Moreover, hydraulic excitation forces are due to the FSI and cause pressure fluctuations, mechanical vibrations, and alternating stresses in different components of centrifugal pump. In recent years, the application of FSI theory to centrifugal pumps became more popular and it is well documented in literature [9–13].

Vibration performance is one of the most important parameters in designing a centrifugal pump. Actually, experimental tests and CFD simulation are the two methods performed in order to obtain the centrifugal pump vibration response. However, both of the two methods cannot be considered in optimizing the vibration performance of the pump using an iterative method. Appropriate metamodels must be established between the decision variables and the concerned objective functions. Therefore, metamodel technique demonstrates its superiority in the optimization problem of engineering.

Kriging metamodels [14] were originally proposed by the South African mining engineer named Danie Gerhardus Krige. With the rapid development of computer technology, Kriging metamodels have been widely used in various fields [15–19]. Kriging metamodel differs from other metamodels because of the optimal unbiased prediction for the unknown response points [20]. Compared to traditionally response surface methods, Kriging shows its superiority in high dimensional nonlinear problems and prediction accuracy due to the stochastic assumption [21], especially for multiobjective optimization problems [22].

This paper presents an effective optimization method based on Kriging metamodel. The presented method optimizes the vibration performance of the centrifugal pump undergoing FSI phenomena, which reasonably take advantages of the FSI simulation, Kriging metamodel and experimental tests. Although considerable researches were devoted to investigating the vibration performance of centrifugal pump, it should be noted that there exists little literature evidence on the vibration optimization of centrifugal pump in particular that combines with FSI phenomenon. The second part of the paper deals with the study of the transient mechanical characteristics based on the optimized centrifugal pump using FSI method.

CENTRIFUGAL PUMP FSI SIMULA-TION MODEL

FSI Governing Equations

In this study, the fluid-structure interaction (FSI) problem's domain Ω consists of two subdomains: Ω_f and Ω_s with the boundaries as Γ_f and Γ_s, respectively. The subscripts f and s denote the fluid part and solid part, respectively. The following section defines the equations that govern the flow and the structural deformations of the pump.

Fluid Flow Equations

The fluid flowing through the centrifugal pump is treated as incompressible and isothermal. For $x \in \Omega_f$, the conservation of mass and Navier-Stokes equations governing the unsteady flow are, respectively, written as

$$\frac{\partial \rho_f}{\partial t} + \nabla \cdot \left(\rho_f \mathbf{v}_f \right) = 0,$$

$$\rho_f \frac{\partial \mathbf{v}_f}{\partial t} + \rho_f \left(\mathbf{v}_f \cdot \nabla \right) \mathbf{v}_f - \nabla \cdot \boldsymbol{\sigma}_f = \mathbf{f}_f,$$

(1)

where ρ_f is the fluid density, $\mathbf{v}_f$ denotes the fluid particle velocity at time , $\mathbf{f}_f$ denotes the body forces per unit of volume on the fluid, and $\boldsymbol{\sigma}_f$ is the stress tensor defined as

$$\boldsymbol{\sigma}_f = -p\mathbf{I} + \mu \left[\nabla \mathbf{v}_f + \left(\nabla \mathbf{v}_f \right)^T \right]$$

(2)

where p is the pressure, $\mathbf{I}$ denotes the unit tensor, and μ represents the absolute viscosity.

Structural Equations

While working, the centrifugal pump undergoes large deformation and rotation. For $x \in \Omega_s$, the conservation of momentum for the solid deformation u_s are described through Lagrangian formulation:

$$\rho_s \frac{\partial^2 \mathbf{u}_s}{\partial t^2} = \nabla \cdot \boldsymbol{\sigma}_s + \mathbf{f}_s,$$

(3)

where ρ_s is the solid density, σ_s represents the Cauchy stress tensor, and denotes the body forces per unit volume on the solid.

Closure for (3) is found by evaluating the stress using the relevant constitutive relations. Moreover, since the centrifugal pump is related to large deformation and rotation, the constitutive equations are described using a stress-strain relationship.

Interaction between Fluid and Solid

As mentioned above, the FSI occurs during the running process of centrifugal pump. Fluid pressure information transfers to the solid, while displacements information of the solid transfers to the flow.

Furthermore, on the no-slip fluid-structural interface, the information exchange between the fluid and solid should follow the equilibrium conditions

$$\mathbf{v} = \frac{\partial \mathbf{u}}{\partial t} \qquad \forall \mathbf{x} \in \Gamma,$$
$$\sigma_f \cdot \mathbf{n} = \sigma_s \cdot \mathbf{n}$$

where Γ denotes the fluid-structural interface and $\Gamma = \Gamma \cap \Gamma_{s,}$, n represents the unit normal at the interface .

Decision Variables

The working process of centrifugal pump involves vibrations, FSI, and energy conversion and loss. As the "heart" component for a centrifugal pump, impeller plays an important role in all these phenomena and transforms the mechanical energy into the kinetic energy of the fluid. Moreover, the geometry shape of impeller blade has strong effect on pump performance, including vibrations. This paper focuses on optimizing the impeller blade to minimize the vibration response of the centrifugal pump.

The recommended number of impeller blades for high head centrifugal pumps is usually between five and seven. In fact, too many blades lead to higher friction losses and may cause low blade loading; fewer blades may result in higher blade loading. Turbulent dissipation losses will rise because of the increased secondary flow and stronger deviation between blade and flow direction. Therefore, six blades are chosen.

Figure 1 shows the main dimensional parameters of the impellers of the centrifugal pump studied in this paper, notably, it is double suction type. In addition, among all kinds of centrifugal pumps, the double suction centrifugal pumps are widely used in industry for various applications due to large flux and high lift.

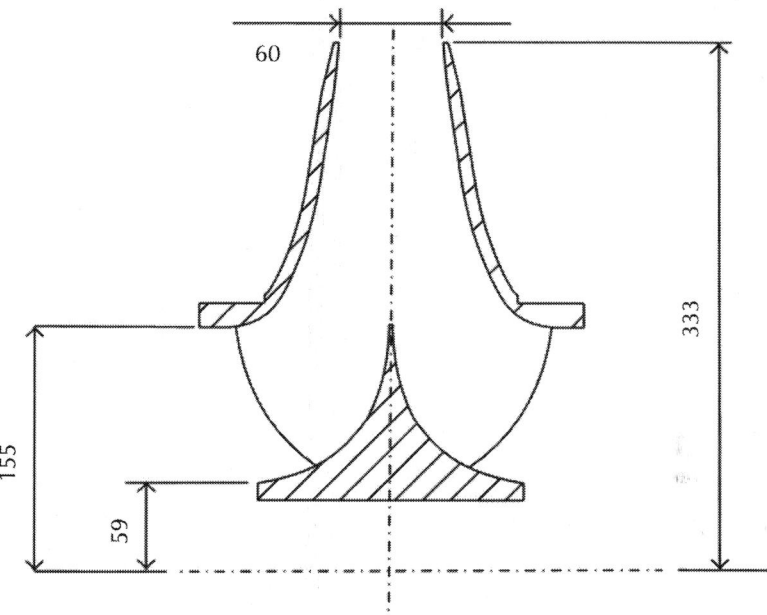

Figure 1: Main dimensions of centrifugal pump's impellers (unit: mm).

Figure 2 shows the meridional section of the impeller. Due to symmetry of the model, the optimization of double suction impeller can be converted into the optimization of a single suction type. Hence, as Figure 2shows, the meridional section of the single suction impeller is actually determined by the solid line and dashed line $\overline{FG}$. The solid line is parameterized by quartic Bézier curve with five control points; the five decision variables for the pump impeller are φ_1, λ_1, φ_2, and λ_2, Table 1 summarizes the boundaries of the decision variables. Here, λ is the relative position in the line segment. For example, taking line segment $\overline{AC}$ λ_1, is the ratio of the line length $\overline{AB}$ to $\overline{AC}$ $l$is the length of line segment $\overline{FG}$

Table 1: Decision variables and their boundaries

Decision variable	Lower boundary	Upper boundary
φ_1 (deg)	0	30

λ_1	0.02	0.98
φ_2 (deg)	70	90
λ_2	0.02	0.98
l (mm)	145	195

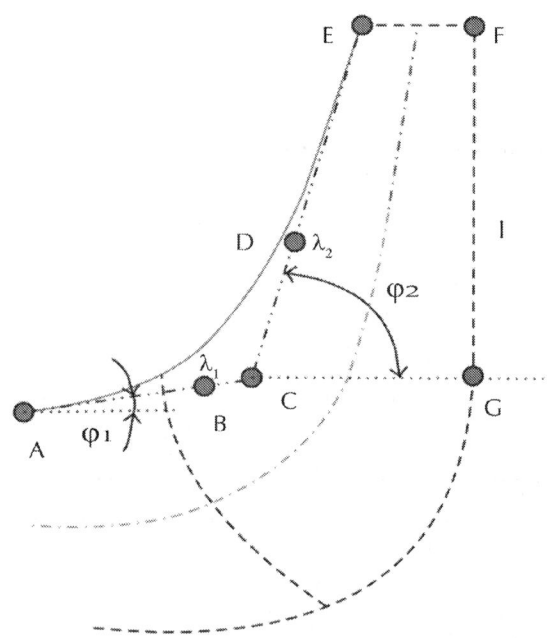

Figure 2: Meridional section of the pump impeller.

FSI Model Sample and Simulation

Latin hypercube sampling (LHS) is a design of experiment (DOE) method originally developed by Mckay in 1979. LHS approach has the space-filling character and can guarantee the sample points covering the entire design domains homogeneously. Hence, 119 simple points and 30 test points have been obtain by LHS method. The simple points are the input data of Kriging surrogate model, while the test points are used to validate the accuracy of the Kriging predictor.

Table 2 summarizes the combinations of decision variables in the sample points. The FSI simulation models are built based on these sample points. Figure 3 shows one case of FSI simulation models. Figure 3(a) corresponds to a full FSI model with solid and fluid parts. Figure 3(b) is the cutaway view of the full FSI model, and Figure 3(c) gives the detailed view of the tongue region. The structural part consists of pump volute casing, impeller, and impeller shaft, while the fluid part is the liquid flowing through the structural part. Moreover, the fluid part is also called the hydraulic model of centrifugal pump.

Table 2: The sample points and corresponding results of FSI simulations

Serial number	Decision variable				(mm)	Objective
	φ_1 (deg)	λ_1	φ_2 (deg)	λ_2		RMS (mm)
1	0.00	0.028	85.76	0.061	157.71	0.5411
2	0.25	0.183	70.00	0.223	183.14	0.4068
3	0.51	0.191	87.29	0.744	184.41	0.4228
4	0.76	0.728	73.90	0.752	185.25	0.3867
5	1.02	0.777	83.22	0.484	161.10	0.4108
6	1.27	0.199	89.66	0.109	156.44	0.5271
7	1.53	0.557	87.80	0.459	189.07	0.5251
8	1.78	0.085	71.19	0.427	168.73	0.5812
9	2.03	0.467	84.24	0.305	161.95	0.5672
10	2.29	0.264	75.08	0.321	155.17	0.4128
11	2.54	0.232	80.51	0.378	147.97	0.3603
12	2.80	0.288	81.53	0.834	177.63	0.4529
117	29.49	0.646	88.64	0.443	163.64	0.3627
118	29.75	0.817	84.75	0.516	168.31	0.3667
119	30.00	0.891	76.10	0.785	189.49	0.5792

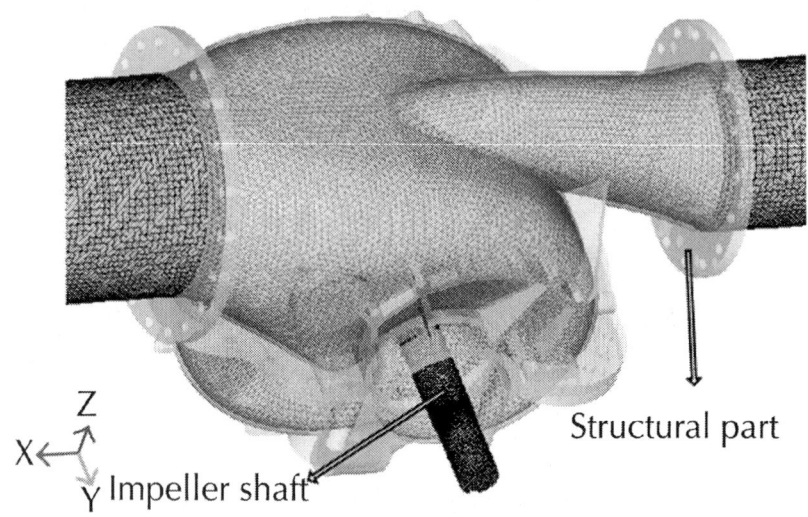

(a)

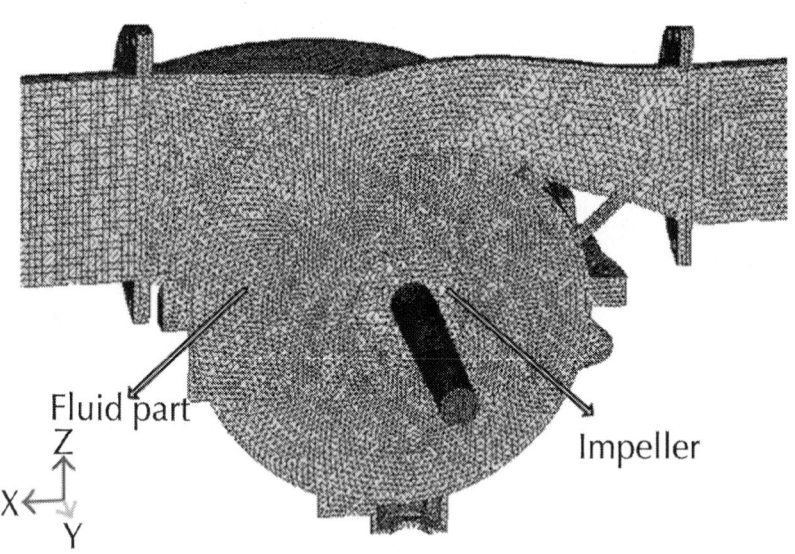

(b)

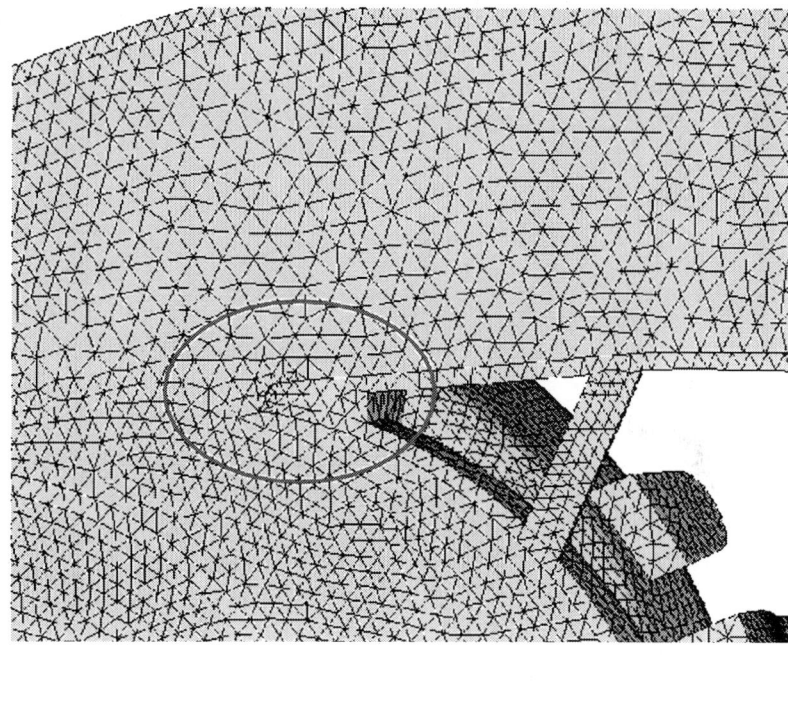

(c)

Figure 3: One case of FSI simulation models.

The calculation of structural part of the pump has been carried out through computational structure dynamics (CSD) analysis, performed using Abaqus FEA software. The pump volute casing and impeller are both made of aluminum-bronze alloy; the elastic modulus is 125000 MPa, the density is 7630 Kg/m³, and the Poisson's ratio is 0.327. The impeller shaft is made of alloy steel, with elastic modulus of 206000 MPa, density of 7800Kg/m³, and Poisson's ratio of 0.3. The increment size of time step is set as, 1×10^{-4} and the total simulation time is 6T s, where T is the cycle of the pump corresponding to a changed angle of 60°. Furthermore, the differential equation of the centrifugal pump at excitation state by FSI can be expressed as

$$\mathbf{M}\ddot{x}(t) + \mathbf{C}\dot{x}(t) + \mathbf{K}x(t) = \mathbf{F}(t),$$

(5)

where t is the time; M, C, and K are the structural mass matrix, structural damping matrix and stiffness matrix, respectively; $\ddot{x}(t)$, $\dot{x}(t)$, and $x(t)$ represent the acceleration vector, velocity vector, and displacement vector, respectively; $F(t)$ denotes the load vector of the node.

The computational fluid dynamics (CFD) has been simulated using fluent code. The fluid is water, with a temperature of 20∘ C, density of 998.2 Kg/m³, and viscosity of 1.003×10^{-3} Pa-s. Table 3 lists the parameters for CFD simulations. The hydraulic models are established by the standard $k - \varepsilon$ turbulence models and wall functions based on logarithmic law, which are consistent with the no-slip condition. Static boundary condition and rotary boundary condition are imposed on the boundary of volute flow domain and impeller flow domain, respectively. Moreover, the interaction between these two boundaries is taken into account through the moving mesh model. The unsteady Reynolds-averaged Navier-Stokes (URANS) equations are calculated by finite volume method (FVM), and the pressure-velocity coupling is solved by means of the SIMPLEC algorithm. Second order upwind discretizations are used for determine and diffusive terms of the turbulence model equations. The residual error is set as 1×10^{-5} to judge whether the calculation is convergent. In addition, the time step and total simulation time are set as 1×10^{-4} and, 6T respectively, in order to correspond with the structural simulation part.

Table 3: Basic parameters for numerical simulations

Parameter Q	Value
Flow rate	2000 (m3/h)
Rotational speed	1400 r/min
Number of blades	6
Inlet operating pressure	1 (atm)

The information exchange between solid (Abaqus) and fluid flow (Fluent) at the coupling interface is performed in the platform of MpCCI. Figure 4 outlines the process of FSI simulation: first, the models of CSD and CFD are prepared independently, such as the setting of material, loads, and boundary conditions. Then, the pressure information of the fluid is transferred to Abaqus for structure analysis; meanwhile, the displacements information of the structure are transferred to Fluent for fluid analysis, and the information exchange at the coupling interface repeatedly until the calculation has converged. Finally, the results of FSI simulation are post processed using both Abaqus and Fluent.

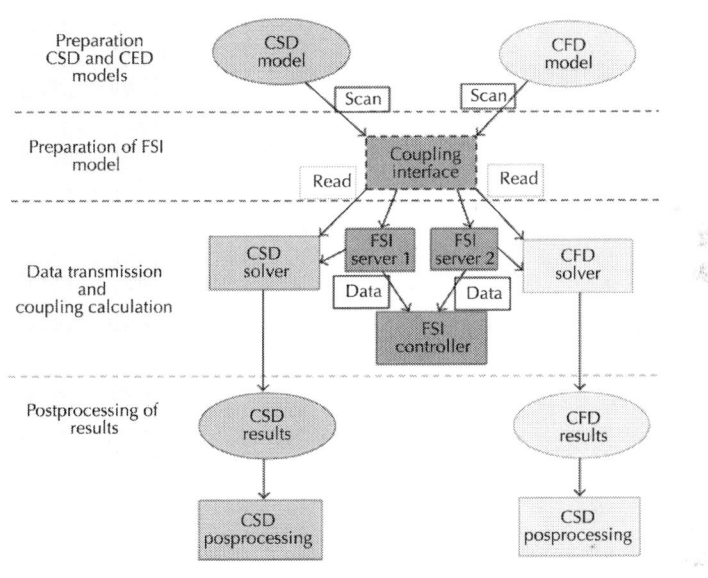

Figure 4: The process of FSI simulation.

As aforementioned, this paper mainly focuses on optimizing the vibration performance of the centrifugal pump using FSI. Hence, the root mean square (RMS) value of the displacement response at the pump bearing block is chosen as the objective function, which can be defined as follows:

$$U_{\mathrm{RMS}} = \sqrt{\frac{1}{N}\sum_{i=1}^{N} U_i^2},$$

(6)

where N is the total number of the time steps, and U_i denotes the displacement response of the th time step, and the direction is the vertical direction of the bearing support. The last column of Table 2 summarizes the results of U_{RMS} calculated through FSI simulations at the sample points, which are the output data used to build the Kriging model.

KRIGING-BASED OPTIMIZATION

Kriging Approach

Kriging predicts unknown values of a random function based on all of the observed points [22]. Moreover, Kriging metamodels show global performance rather than local characteristics. A combination of a global model and localized departures of the form describes a Kriging model; thus

$$y(x) = \sum_{k=1}^{n} \beta_k f_k(x) + Z(x),$$

(7)

where $y(x)$ is the response function, , $f(x) = [f_1(x), \ldots, f_n(x)]^T$ is the regression basis function, n is the number of the basis function, and $\beta = [\beta_1,...,\beta_n]^T$ is the regression coefficient. is assumed as a realization of an independent Gaussian random process with zero mean and spatial correlation function given by [23]

$$Cov[Z(\tau), Z(x)] = \sigma^2 R(\theta, \tau, x),$$

(8)

where σ^2 denotes the process variance, $R(\theta, \tau, x)$ is the correlation function between the points τ and, x and θ is the unknown correlation parameter. Several types of correlation models, such as linear correlation model and exponential correlation model can be considered. However, the Gauss correlation model adopted in this paper is more popular in Kriging met models with the form

$$R(\theta, \tau, x) = \exp\left(-\sum_{j=1}^{m} \theta_j (\tau_j - x_j)^2\right),$$

(9)

where the quantities τ_j and x_j, respectively, denote the jth components of sample points τ and; x;m is the dimension of the decision variables.

The predicted value and estimation error at point x are, respectively, given by

$$\hat{y}(x) = \mathbf{f}^T(x)\hat{\beta} + \mathbf{r}^T(x)\mathbf{R}^{-1}(\mathbf{Y} - \mathbf{F}\hat{\beta}),$$

$$s(x) = \sigma^2\left(1 + \mathbf{u}^T(\mathbf{F}^T\mathbf{R}^{-1}\mathbf{F})^{-1}\mathbf{u} - \mathbf{r}(x)^T\mathbf{R}^{-1}\mathbf{r}(x)\right),$$

(10)

where Y represents the response of the sample points, , u = $F^TR^{-1}r(x)$ − f(x), F, is a vector which is composed by the value of at each sample point, denotes a vector which represents the correlation between an unknown point and all known sample points. In addition, r (x) = [$(\theta,$ $x, x1)$ · · · $R(\theta, x, xN)$], N is the total number of the sample points, is an symmetric correlation matrix written in the following form:

$$\mathbf{R} = \begin{bmatrix} R(x_1, x_1) & \cdots & R(x_1, x_N) \\ \vdots & \ddots & \vdots \\ R(x_N, x_1) & \cdots & R(x_N, x_N) \end{bmatrix}.$$

(11)

Under the unbiased condition, the unknown parameters β and σ^2 can be estimated through

$$\hat{\beta} = (\mathbf{F}^T\mathbf{R}^{-1}\mathbf{F})^{-1}\mathbf{F}^T\mathbf{R}^{-1}\mathbf{Y},$$

$$\hat{\sigma}^2 = \frac{1}{m}(\mathbf{Y} - \mathbf{F}\hat{\beta})^T\mathbf{R}^{-1}(\mathbf{Y} - \mathbf{F}\hat{\beta}).$$

(12)

As a matter of fact, once the types of regression model and correlation model have been chosen, the correlation matrix R and unknown parameters β and σ^2 all depend on the correlation parameter θ. Thus, a Kriging metamodel is completely established only if the value of θ is determined. Furthermore, the most commonly used approach to calculate the value of correlation parameter θ is maximum likelihood estimation (MLE), and the problem can be converted into an unconstrained global optimization problem as follows [20]:

$$\text{Minimize} \quad \left\{ \psi(\boldsymbol{\theta}) = \sigma(\boldsymbol{\theta})^2 \left| \mathbf{R}_{\theta} \right|^{1/m} \right\}$$

$$\text{Subject to} \quad \boldsymbol{\theta} > 0. \tag{13}$$

Modeling and Verification of the Kriging Model

Kriging meta model is established according to Table 2.The input data is the set of 119 sample points obtained through LHS method, and the input variables are the impeller geometric parameters φ_1, λ_1, φ_2, λ_2, and l shown in Figure 2.The output data are the results of FSI simulations corresponding to the sample points, and the output variable is the RMS value of the displacement response URMS. Table 4 shows the parameters of the Kriging meta model, β, σ^2, and θ

Table 4: The parameters of the Kriging model

Parameter	Value
β	$2.688_e 5$, −0.142, −0.062, −0.036, −0.093, 0.107
σ^2	0.00453
θ	8.406, 14.718, 18.851, 21.512, 21.401

The Kriging meta model can be applied to the vibration optimization only if the Kriging predictor's estimated accuracy is higher enough. Otherwise, the meta model should be rebuilt by adjusting the parameters. An additional set of 30 points obtained through LHS method is used as test points to verify the performance of Kriging's predictor.

The FSI analysis gives the RMS values of the displacement response corresponding with the test points. In addition, for the FSI simulations based on the test points, the basic parameters and boundary conditions are the same with the sample points.

Figure 5 shows the results of the displacement response's RMS values obtained by Kriging predictor and FSI simulations at the test points. Results show that the predicted values of the Kriging meta model correspond to the FSI simulation values. Hence, the vibration optimization of centrifugal pump can be performed based on the Kriging surrogate model.

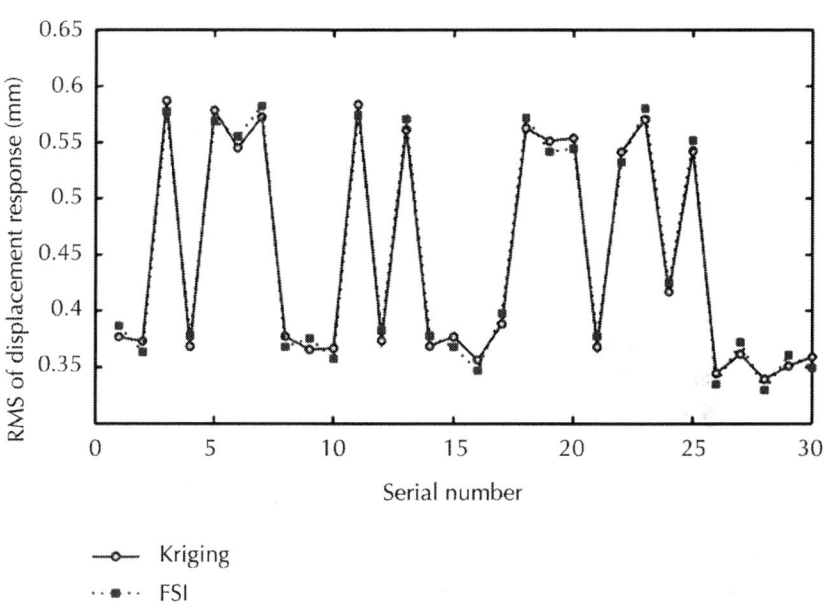

Figure 5: The results of RMS values at test points.

Optimization Based on the Surrogate Model

The optimization problem of centrifugal pump in this paper can be given as follows:

Find $\mathbf{X} = \left[\varphi_1, \lambda_1, \varphi_2, \lambda_2, l \right]^T$

Minimize $U_{RMS} = f\left(\varphi_1, \lambda_1, \varphi_2, \lambda_2, l \right)$

Subject to $0° \leq \varphi_1 \leq 30°$

$0.02 \leq \lambda_1 \leq 0.98$

$70° \leq \varphi_2 \leq 90°$

$0.02 \leq \lambda_2 \leq 0.98$

$145\,\text{mm} \leq l \leq 195\,\text{mm},$

(14)

where is f the Kriging approximation of the displacement response's RMS values.

The above-defined problem can be resolved through multi-island genetic algorithm (MIGA), a modified version of genetic algorithm (GA). MIGA decomposes the population in one generation into several subpopulations. The subpopulations are also called "Islands," and the genetic operations are executed on each "Island" independently. Furthermore, this independency can prevent the optimization solution from local optima. Table 5 lists the detailed parameter settings used for MIGA. The experiments are carried out on a Desktop PC with Intel Core 2 quad CPU and 3.25 GB RAM; all of the cores have the speed of 2.66 GHz. Due to stochastic behavior of MIGA algorithm, at least 30 independent runs are required to provide the results with statistical confidence.

Table 5: The parameter settings of MIGA

Parameters	Value
Size of subpopulation	100
Number of islands	10
Number of generations	10
Gene size	32
Rate of crossover	1.0
Rate of mutation	0.01
Rate of migration	0.5

Interval of migration	5
Number of runs for the problem	30

RESULT AND DISCUSSION

Optimization Result and Validation

Table 6 shows the optimization result and average CPU time. The RMS value of the displacement response improves to 0.3341 mm. Moreover, the accuracies of Kriging meta model and FSI simulation have been further validated through experimental tests. Thus, a prototype of centrifugal pump based on the geometric parameters in Table 6 has been produced, as shown in Figure 6. The pump is fixed on the test bench; a 1 MW electric machinery drives the pump impeller. A displacement sensor installed at the pump bearing measures the displacement response of the bearing block. The water circulates in a close loop and the flow rate is constant. The basic test parameters correspond to parameters of FSI simulations summarized in Table3.

Table 6: The result of optimization

1 (deg)	1	2 (deg)	2	(mm)	RMS (mm)	Average time (s)
26.37	0.938	83.31	0.934	156.89	0.3341	1836

Figure 6: The prototype of centrifugal pump corresponding to the optimization result.

Table 7 compares the results of Kriging predictor, FSI simulation, and experimental test. The results given by the three methods well agree to each other. The error of Kriging metamodel is 3.1%, the error of FSI simulation is 4.4%, and they are both less than 5%. It is well known that the experimental test plays an indispensable role in validating the optimization design for centrifugal pump. However, manufacturing a prototype pump or the experimental equipment is expensive. Furthermore, due to the complexity of the model, the FSI simulation is time-consuming. Hence, the optimized design of the pump should minimize both elements: costs and calculation time.

Table 7: Results of Kriging, FSI simulation, and experiment

	Kriging	FSI	Experiment
RMS (mm)	0.3341	0.3296	0.3447

This research shows that the predictive ability of the Kriging model has been well justified both by FSI simulations and by experimental test. Therefore, the well validated surrogate model can completely replace time-consuming FSI simulations and substitute a great majority of expensive experiment tests. That is, the Kriging surrogate model provides great convenience in studying the vibration performance of centrifugal pump, especially for accumulating the practical experience of pump design. Moreover, the well validated surrogate model can benefit both the further development of centrifugal pump manufacturer and the improvement of the pump designer's ability. Therefore, the surrogate model method makes the investigation of pump performance easy, which is of course on the promise that the model accuracy is high enough.

The Analysis of Impeller Mechanical Behavior through FSI

The mechanical characteristics of the pump impeller are significant for the working behaviors of centrifugal pumps. During the working process of a centrifugal pump, the periodic hydraulic loads imposed on the pump will lead to the dynamic deformation of the impeller and impeller shaft. Moreover, the dynamic deformation will further influence the flow field distribution. The analysis of the mechanical behavior of the impeller is a typical FSI problem. In general, there are mainly three types of loads acting on pump impeller: coupling pressure load from the fluid, gravity, and inertia force due to the circular motion. However, all these loads are finally balanced by the support reaction of the bearings and the input moment of the pump. FSI method allows investigating the dynamic force of the impeller and the input moment, and the calculation results are important to highlight the mechanical properties of the centrifugal pump. For example, the analysis results can help in choosing the appropriate sizes and types of impeller shaft and bearings.

Actually, either the radial force of the pump impeller or the input moment of the pump cannot be easily measured because of the expensive measuring equipment and complex multipoints installation. When the simulation model is accurate, FSI simulation method shows advantage in obtaining the radial force and input moment. Furthermore, as previously mentioned in Section 4.1, the results obtained by Kriging predictor's, FSI simulation, and experiment test well agree to each other. The comparison indicates that the FSI simulation model is well validated, and FSI simulation model leads to reliable results. This research investigates the radial force of the pump impeller and the input moment of the pump through FSI method. In addition, the radial force and input moment are calculated based on the FSI simulation model in Section4.1. The basic settings of the FSI simulation are unchanged, such as the definitions of material properties, simulation time step, boundary condition, and coupling interface.

Figures 7 and 8 show the results of dynamic radial force of pump impeller and dynamic input moment of the pump, respectively. The time-dependent transient force and moment in each time step are calculated by direct integration method. Both the curve of radial force and moment indicate cyclical fluctuation in general and six cycles corresponding to a full pump impeller revolution. However, as a result of the tongue region shown in Figure 3(c), there exists more or less disturbance on the wave crest or wave trough.

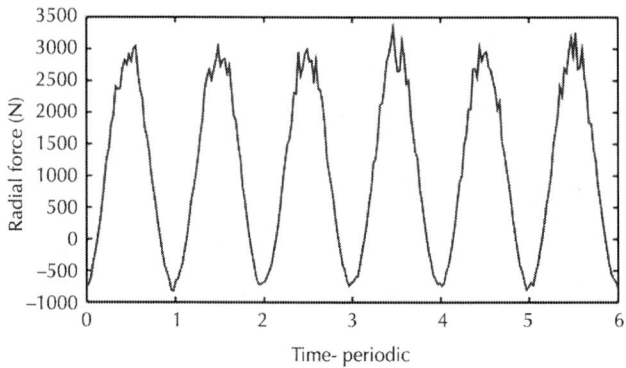

Figure 7: The radial force of the pump impeller.

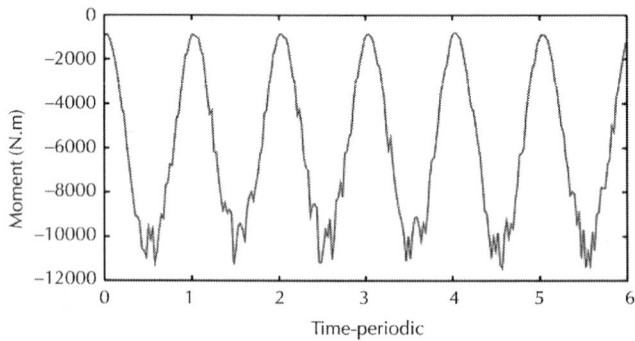

Figure 8: The input moment of the pump.

CONCLUSIONS

This paper proposes a Kriging-based optimization method for the vibrations optimization of centrifugal pumps, which well integrates Kriging surrogate model, FSI simulations, and experimental tests. Moreover, the proposed method overcomes the faults of expensive computation and cost, and it has been proved to be effective on improving pump vibration performance in terms of minimum cost and reduction of development period.

The Kriging surrogate model of pump vibration performance has been established based on the sample points, and the results at the test points showed that the Kriging predictor well agreed with the FSI simulations. The final optimized decision variables have been obtained using MIGA; a prototype has been manufactured according to optimized values of geometrical parameters of the pump. Experimental tests carried out on prototype well agreed with the results of Kriging metamodel and FSI simulation.

Furthermore, based on the final optimized decision variables, the dynamic mechanical performance of pump impeller was further investigated using FSI method. The results showed that the radial force curve and moment curve exhibited cyclical fluctuation.

ACKNOWLEDGMENT

This research was supported by the National Natural Science Foundation of China (no. 11172108). This financial support is gratefully acknowledged.

REFERENCES

1.   J. S. Anagnostopoulos, "A fast numerical method for flow analysis and blade design in centrifugal pump impellers," Computers and Fluids, vol. 38, no. 2, pp. 284–289, 2009. View at Publisher · View at Google Scholar · View at Zentralblatt MATH

2.   L. Zhou, W. Shi, and S. Wu, "Performance optimization in a centrifugal pump impeller by orthogonal experiment and

numerical simulation," Advances in Mechanical Engineering, vol. 2013, Article ID 385809, 7 pages, 2013.

3.   S. Derakhshan, B. Mohammadi, and A. Nourbakhsh, "The comparison of incomplete sensitivities and Genetic algorithms applications in 3D radial turbomachinery blade optimization," Computers and Fluids, vol. 39, no. 10, pp. 2022–2029, 2010.

4.   A. Papierski and A. Blaszczyk, "Multiobjective optimization of the semi-open impeller in a centrifugal pump by a multilevel method," Journal of Theoretical and Applied Mechanics, vol. 49, no. 2, pp. 327–341, 2011.

5.   M. R. Hodkiewicz and M. P. Norton, "The effect of change in flow rate on the vibration of double-suction centrifugal pumps," Proceedings of the Institution of Mechanical Engineers E: Journal of Process Mechanical Engineering, vol. 216, no. 1, pp. 47–58, 2002.

6.   S. Guo and Y. Maruta, "Experimental investigations on pressure fluctuations and vibration of the impeller in a centrifugal pump with vaned diffusers," International Journal B, vol. 48, no. 1, pp. 136–143, 2005.

7.   C. G. Rodriguez, E. Egusquiza, and I. F. Santos, "Frequencies in the vibration induced by the rotor stator interaction in a centrifugal pump turbine," Journal of Fluids Engineering, vol. 129, no. 11, pp. 1428–1435, 2007.

8.   F. J. Wang, L. X. Qu, L. Y. He, and J. Y. Gao, "Evaluation of flow-induced dynamic stress and vibration of volute casing for a large-scale double-suction centrifugal pump," Mathematical Problems in Engineering, vol. 2013, Article ID 764812, 9 pages, 2013.

9.   F. K. Benra, "Application of fluid/structure interaction methods to determine the impeller orbit curves of a centrifugal pump," in Proceedings of the 5th IASME/WSEAS International Conference on Fluid Mechanics and Aerodynamics, pp. 169–174, Athens, Greece, 2007.

10.   A. Fontanals García, A. D. J. Guardo Zabaleta, M. G. Coussirat Núñez, and E. Egusquiza Estévez, "Numerical study of the fluid—structure interaction in the diffuser passage of a centrifugal pump," inProceedings of the 4th International Conference on Computational Methods for Coupled Problems in Science and Engineering, pp. 1–10, 2011.

11. Q. Jiang, L. Zhai, L. Wang, and D. Wu, "Fluid-structure interaction analysis on turbulent annular seals of centrifugal pumps during transient process," Mathematical Problems in Engineering, Article ID 929574, Art. ID 929574, 22 pages, 2011.

12. S. Yuan, J. Pei, and J. Yuan, "Numerical investigation on fluid structure interaction considering rotor deformation for a centrifugal pump," Chinese Journal of Mechanical Engineering, vol. 24, no. 4, pp. 539–545, 2011.

13. J. Pei, S. Yuan, and J. Yuan, "Fluid-structure coupling effects on periodically transient flow of a single-blade sewage centrifugal pump," Journal of Mechanical Science and Technology, vol. 27, no. 7, pp. 2015–2023, 2013.

14. J. P. C. Kleijnen, "Kriging metamodeling in simulation: a review," European Journal of Operational Research, vol. 192, no. 3, pp. 707–716, 2009.

15. T. W. Simpson, T. M. Mauery, J. J. Korte, and F. Mistree, "Kriging models for global approximation in simulation-based multidisciplinary design optimization," AIAA Journal, vol. 39, no. 12, pp. 2233–2241, 2001.

16. S. Sakata, F. Ashida, and M. Zako, "Structural optimizatiion using Kriging approximation," Computer Methods in Applied Mechanics and Engineering, vol. 192, no. 7-8, pp. 923–939, 2003.

17. D. Huang, T. T. Allen, W. I. Notz, and N. Zeng, "Global optimization of stochastic black-box systems via sequential kriging meta-models," Journal of Global Optimization, vol. 34, no. 3, pp. 441–466, 2006.

18. E. Davis and M. Ierapetritou, "A kriging based method for the solution of mixed-integer nonlinear programs containing black-box functions," Journal of Global Optimization, vol. 43, no. 2-3, pp. 191–205, 2009.

19. H. Li, T. Qiu, B. Zhu, J. Wu, and X. Wang, "Design optimization of coronary stent based on finite element models," The Scientific World Journal, vol. 2013, Article ID 630243, 10 pages, 2013.

20. J. D. Martin, "Computational improvements to estimating Kriging metamodel parameters," Journal of Mechanical Design, vol. 131, no. 8, Article ID 084501, 7 pages, 2009.

21. A. Matta, M. Pezzoni, and Q. Semeraro, "A Kriging-based algorithm to optimize production systems approximated by analytical models," Journal of Intelligent Manufacturing, vol. 23, no. 3, pp. 587–597, 2012.

22. M. Zakerifar, W. E. Biles, and G. W. Evans, "Kriging metamodeling in multiple-objective simulation optimization," Simulation, vol. 87, no. 10, pp. 843–856, 2011.

23. R. Jin, W. Chen, and T. W. Simpson, "Comparative studies of metamodelling techniques under multiple modelling criteria," Structural and Multidisciplinary Optimization, vol. 23, no. 1, pp. 1–13, 2001.

Design Mixers to Minimize Effects of Erosion and Corrosion Erosion

Julian Fasano[1], Eric E. Janz[2], and Kevin Myers[3]

[1]Mixer Engineering Co., 2673 Stonebridge Drive, Troy, OH 45373, USA

[2]Chemineer, Inc., 5870 Poe Avenue, Dayton, OH 45414, USA

[3]Department of Chemical & Materials Engineering, University of Dayton, 300 College Park, Dayton, OH 45469-0246, USA

ABSTRACT

A thorough review of the major parameters that affect solid-liquid slurry wear on impellers and techniques for minimizing wear is presented. These major parameters include (i) chemical environment, (ii) hardness of solids, (iii) density of solids, (iv) percent solids, (v) shape of solids, (vi) fluid regime (turbulent, transitional, or laminar), (vii) hardness of the mixer's wetted parts, (viii) hydraulic efficiency of the impeller (kinetic energy dissipation rates near the impeller blades), (ix) impact velocity, and (x) impact frequency. Techniques for minimizing the wear on impellers cover the choice of impeller, size and speed of the impeller, alloy selection, and surface coating or coverings. An example is provided as well as an assessment of the approximate life improvement.

INTRODUCTION

There are numerous applications of mixers that deal with erosive solids, especially in the minerals processing and power industries. In many of these applications, there is an erosion-corrosion synergistic effect on the wear of a mixer's wetted parts, particularly the impeller. This paper pulls together the authors' research with numerous articles on erosion and erosion corrosion to permit a designer to optimize the cost-based life of eroding mixer parts before replacement is required.

There are a large number of factors that can affect the rate of erosion. Many of these factors have been known and studied to some extent:

- chemical environment,
- hardness of solids,
- density of solids,
- difference in liquid and solid density,
- percent solids,
- shape of solids,
- fluid regime (turbulent, transitional, or laminar),
- fluid rheology (e.g., pseudoplasticity),
- hardness of the mixer's wetted parts,
- young's modulus of the mixer's wetted parts,
- hydraulic efficiency of the impeller (kinetic energy dissipation rates near the impeller blades),
- impact velocity,
- impact frequency,
- angle of impact.

Theoretically the rate of volume loss of material is due to the kinetic energy lost when a particle impacts a material [1]. This would suggest a velocity exponent of 2. However, presented below, experimental velocity exponents have ranged from 1.5 to 4.0. The general form of the equation relating erosion rate to velocity is given by

$$E = K \cdot V^n f(\theta),$$

(1)

where E= volumetric erosion rate, K= constant (function of all parameters other than V or θ), V= particle velocity or relative velocity for rotating systems (impellers), n= velocity exponent (can also be a function of other parameters), and θ = impingement angle.

Most investigators have used this general equation form.

Sapate and RamaRao [2] used a power law correlation between volumetric erosion rate and impingement velocity in a nonrotating system. They observed exponents on velocity of 1.91 to 2.52. The velocity exponent showed an increasing trend with increasing hardness of the alloys irrespective of the hardness of the erodent particles and the impingement angle of the alloys investigated.

Stack [3] and others investigated the effect that corrosion plays in an erosion environment. These investigators studied various parameters of the corrosion-erosion environment in a nonrotating system. They observed velocity exponents that ranged from 1.4 to 3.5. They concluded that exponents derived for erosion of alloys under erosion-dominated conditions can be correlated to those derived for the strictly ductile erosion process. These are typically very near the theoretical "2" for the strictly ductile erosion process. However, those for the erosion corrosion dominated regime are higher than for the erosion-dominated regime and were in the 2.5 to 3.5 range. A publication by the Hydraulics Institute [4] suggests that the erosion velocity exponent for pumps in slurry transport is on the order of 2.5–3.0.

Fort [5] and others studied pitched blade impellers 100 mm in diameter with a blade width of 20 mm in water-solid slurries under turbulent conditions. These impellers were studied at pitch angles of 20, 35, and 45. These studies included a slurry of 18.3% by volume of gypsum having a mean particle diameter of 0.1 mm and a 10% by volume slurry of 0.4 mm mean diameter sand particles. From their studies, they concluded the following.

Particles of the lower hardness gypsum generated uniform sheet erosion over the entire surface of the impeller, while the particles of sand, having a higher hardness, generated predominately erosion of the leading edges of the impeller blade.

- (The higher the hardness of the blade material, the lower the wear rate of the blade.
- The wear rate of the leading edge was not a function of pitch angle.
- Sheet erosion of the blades exhibits a maximum erosion rate between 20 and 45 .

Zheng [6] and others studied the erosion-corrosion synergistic effect in an acidic slurry. The slurry was 10% by weight H_2SO_4 and 15% by weight −60 mesh (<0.251 mm) corundum sand. Their apparatus was a rotating disk with four specimen holders on its edge. They determined the rate of erosion by making electrochemical measurements during rotation. All of their studies were done under turbulent flow conditions. Erosion rate velocity exponents ranged from 1.9 to 4.0. A model was proposed and used which divided the overall erosion rate into an erosion rate via corrosion, an erosion rate via erosion, and an erosion rate due to synergism. The synergism rate was very large and varied between 32 and 99% of the total. The percent contributed by synergism diminished as the alloy became more statically corrosion resistant.

Amelyushkin and Agafonov [7] studied the erosion of cogeneration steam turbine blades caused by water droplets. If kinetic energy is high enough, even water droplets can cause erosion. They found that the toroidal and near root vortices were very intense and caused enhanced wear of the rotor blades. Also they found that they were able to eliminate erosion by making the water droplets small enough. It is expected that these effects are related to grain size. In ductile erosion, plastic deformation may occur first, before metal is removed. If erosion is due to intergranular grain fracture, then if particles are significantly smaller than the metal's grain size, erosion should be minimized. As ductile alloy grain sizes are on the order of 20 μm, particles smaller than this should have little erosive effect.

Khalid and Sapuan [8] studied wear for a centrifugal pump impeller in a slurry application. Weight and diameter losses were very nearly linear with time over 480 hours of operation. Blade height and depth loss did exhibit some nonlinearity but were modeled as linear. Typical of rotating devices in slurries, more material was lost near the periphery of the impeller than in the center because linear velocities increase with radial distance from the shaft.

López [9] and others studied the effect of corrosion erosion at relatively high velocities on 304 and 420 stainless steel. Velocities ranged from 4.5 to 8.5 m/s. Such velocities are not common in mixing equipment except in high-shear devices. They used a rotating disc device with erosion samples attached to the periphery of the disc. The aqueous liquid for the slurry was composed of 70% by weight H_2SO_4 and 3.5% NaCl. The slurry solids were 30% by weight SiO_2 particles with a mean diameter about 0.25 mm. They found that high-velocity impacts were beneficial. The combined action of erosive and corrosive mechanisms did not lead to a significant increase in mass loss if compared to corrosion tests. They suggested that pores, small cracks, and fresh pits can be covered by the prows and lips that are formed as a consequence of the wedge action of round particles which produces a smoother, uniformly corroded surface. Thus, even though the transport mechanisms which remove corrosion products are greater due to higher velocities, the surface area exposed is smaller.

Corpstein and Fasano [10] studied slurry wear through multilayer paint modeling. The paint layers are on the order of 0.0381 mm thick. The three layers of paint used had an overall thickness of approximately 0.114 mm. This is only about 7% of the blade thickness and did not change the fluid hydraulics over the blade. Erosion was studied using 8.3-inch diameter scaled down axial flow mixing impellers in a sand water slurry. These studies pointed out the strong effects that blade-shedding vortices, could have on erosive wear. Impellers, such as the four-bladed-pitched impeller that generated stronger vortices, suffered the highest degree of localized erosion. The effects of these vortices can completely wear through an impeller blade, leaving holes where the vortices contacted the surface. High-efficiency impeller blades created significantly smaller vortices and as a consequence exhibited much lower localized wear. Vortex erosion can be severe and occurs on the backside (low-pressure side) of the impeller blade. Comparisons of the backside wear pattern for a Chemineer HE-3 impeller and a standard generic 45° pitched four-bladed impeller are shown in Figures 1 and 2. The impellers were each 211 mm diameter and operated at 870 rpm in a 10% by weight sand slurry in water. The weight mean particle size of the sand was 360 μm. These backside erosion patterns were made after only 30 minutes of operation, and it is obvious that the erosion on the backside of the 45° four-bladed-pitched impeller was much more severe than the erosion for the Chemineer HE-3 high-efficiency impeller.

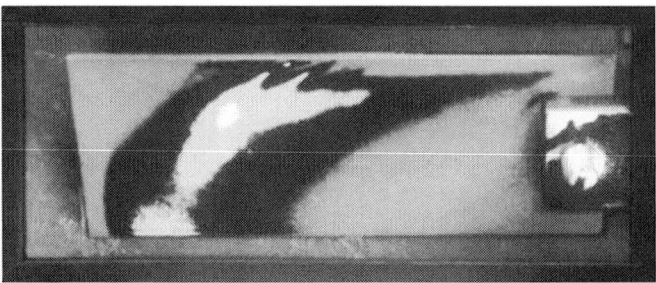

Figure 1: Backside of pitched blade impeller.

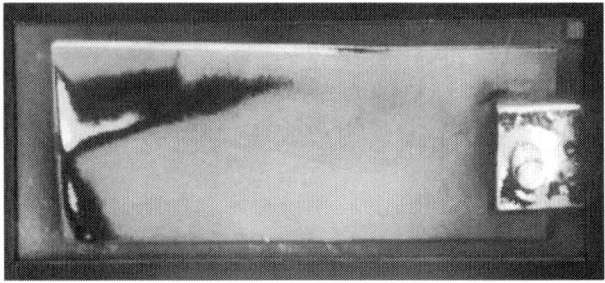

Figure 2: Backside of HE-3 high-efficiency impeller.

Wu [11] and others were also successful in using this technique to study five different style radial flow impellers and a low attack angle (~15°) (6-bladed) pitched impeller. As expected, the hydraulically more efficient six-bladed-pitched impeller experienced the least erosion.

Increased hardness of metals will generally provide an increased life. Miller and Schmidt [12] compared the erosion rates of 16 metals in a recycled slurry test system using 2% by weight silica sand in water. The impeller tip velocity was 15.7 m/s, and the temperature was 16 C. In addition to the erosion rate for each metal, they included the metal's hardness. The best fit for their data was logarithmic. However, probably due to the synergistic corrosion effects, the data was fairly dispersed. A plot of this data is provided in Figure 3. The effect of particle hardness depends on whether erosion is ductile or brittle. For brittle erosion, the effect of particle hardness is much more pronounced than for ductile erosion.

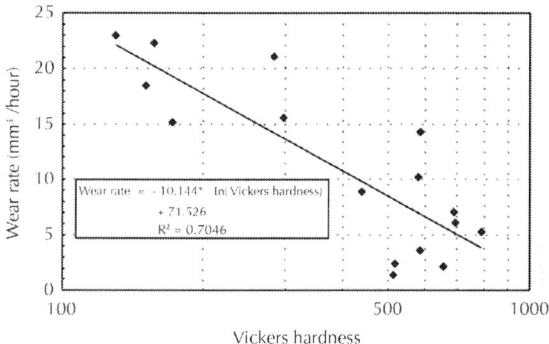

Figure 3: Wear rate data of Miller and Schmidt.

Changes in particle size can change the erosion mechanism. Stachowiak and Batchelor [13] reported that as the particle size was increased from 8.75 μm to 127 μm, the mode of erosion changed from ductile to brittle. The erosion study was for silicon carbide particles impinging on glass, steel, graphite, and ceramics. The particle velocity was 152 m/s.

DESIGN FOR EROSION MINIMIZATION

Because maximum velocities in mixing processes seldom exceed 6 m/s, erosion and corrosion erosion of materials are fatigue processes for most mixing processes. There is generally not enough particle kinetic energy to cause ductile erosion where there is some plastic flow of material. The fatigue process occurs on a micro- or localized scale, and, as with macroscale fatigue, two stages of the erosion process have been observed. There is an incubation period followed by the formation and growth of pits involving the removal of the metal or material. Refer to a materials behavior text such as that by Armstrong and Zerilli [14] for a more in-depth discussion on material behavior.

Due to the vast number of parameters that can affect erosion or erosion-corrosion processes, and the fact that this area of mixer service has not been widely studied, it is very difficult to predict a priori what the rate of erosion will be for any given liquid-solid application.

However, there are certain factors within the control of the designer that can be used to optimize the life of the mixer's wetted parts.

Most mixer designers will not have control over the type of slurry, the percent solids, the hardness of the solids, the shape of the solids, the liquid, the pH, and so forth. However designers will generally have control over

- the mixer's wetted parts materials, coating, or lining,
- the impeller style,
- the impeller horsepower and speed combination.

Material Selection

The choice in selecting a material is to either go hard or soft and elastic. All else being equal, when selecting a metal alloy, a higher hardness will lead to a longer life. Thus, when selecting a metal alloy material, select a hard material which will also provide good corrosion resistance.

There are a number of hard surface ceramic coatings such as tungsten carbide or silicon carbide, which could be applied to the high-wear areas such as impeller blades. Ceramics are the most wear resistant but are low in toughness and impact strength. Ceramic coatings as well must be corrosion resistant to the liquid medium. Ceramics also do not have the ability to absorb much strain. These high strains on flexing blades may allow cracking of the ceramic coating. Ceramic coating applicators should be able to provide the maximum allowable strain for the ceramic coating under consideration. Coatings of the more common ceramic materials tend to be more costly than high-hardness metals, or elastomeric coverings [15].

Glass-lined equipment has a glass hardness of 5 to 6 on the Moh scale. For the great majority of solids, this hardness would be very acceptable. However, there are numerous materials and minerals including, Al_2O_3, SiO_2, WC, SiC, and ZiO_2 that have higher harnesses and would tend to wear away the glass lining. Glass linings have very many of the same limitations as ceramic coatings. They tend to be brittle and cannot tolerate much strain.

Elastomeric coverings on the order of 3/8" thick for industrial scale impellers have a long history of providing longer life in slurry applications. Instead of having to absorb most of the particle's impact energy, an elastomer releases most of the energy back to the particle after impact. Elastomeric lining manufacturers and applicators will generally recommend an elastomeric hardness of 40–60 Durometer A for optimum life. As with metals or hard surface coatings, the lining must also be compatible with the fluid medium. An elastomer's hardness is directly related to its corrosion resistance. However, as an elastomer's hardness increases above a 40 A Durometer, its erosion resistance decreases. A Durometer selection of 40–60 A is somewhat of a compromise between erosion and corrosion resistance. Elastomers should not be used when large particles are present. The term "large particles" is relative to the impinging velocity and mass of the particle, as well as the thickness of the elastomeric covering. If the impinging particle can bottom out against the metallic substrate, elastomeric coverings should not be used. Even if most of the slurry might be suitable, a small percentage of tramp particles can do significant damage to the elastomeric covering. Since impact energy is a function of the impingement angle, the leading edges of impeller blades are almost always double wrapped. The most popular elastomeric coverings are natural rubber, neoprene, butyl, chlorobutyl, and hypalon. Improperly applied linings on high-efficiency impellers that significantly change the profile of the blade can cause increased erosion problems. Typically linings are double layered on the leading, trailing, and outside edges of impeller blades. These linings must be adequately feathered such that the transition from the double layer to the single layer is smooth to avoid generation of additional vortices.

Both thermoplastic and thermoset polymers do not have the ability to restore back to the particle most of the kinetic energy and are generally not as good in mixing slurry service. Hercules 1900 UHMWPE, touted as being a very abrasion-resistant polymer, was tested by the authors against polymeric protective coatings elastomers 2001-B (natural rubber Durometer A 30–40) and 1054-B (chlorobutyl rubber Durometer A 35–45). The testing procedure was identical to that specified in the Hercules 1900 UHMWPE bulletin. A 50% sand-water slurry was used with a sand weight mean particle size of 53 μm. The specimen tip speed was 2.22 m/s. Weight loss was determined at various time intervals over an 8-hour period. The weight loss versus time was found to be linear with R^2 values for all three falling between 0.93 and 0.94. The rates of erosion were shown in Table 1.

Table 1:

Material	Rate of weight loss, g/hr
Hercules 1900 UHMWPE	0.0975
Polymeric protective coatings 2001-B (natural rubber)	0.0104
Polymeric protective coatings 1054-B (chlorobutyl)	0.0124

As can be observed, the rate of weight loss for the thermoplastic polymer is 7 to 10 times greater than that for the elastomers tested

Dickey and Fasano have provided a general reference on materials selection considerations [16].

Impeller Selection

There are many different impeller styles available to the designer. Selecting the correct impeller can often make a difference in impeller life of two or three times. Erosion of impeller blades can depend heavily on the flow regime, with flow regime being determined by the impeller Reynolds number

$$N_{Re} = \frac{\rho N D^2}{\mu},$$

(2)

where p = density, N = impeller rotational speed, D = impeller diameter, μ = viscosity, and N_{Re} = Reynolds number, dimensionless.

Impellers in turbulent flow create shedding vortices that attach themselves to the back of impeller blades. There are a number of techniques that can be used to visualize these vortices. In Figures 4 and 5, telltales attached to the blade are used to visualize these vortices [17]. They are shown for both a relatively inefficient 45° pitched four-blade impeller and the Chemineer HE-3 high-efficiency impeller.

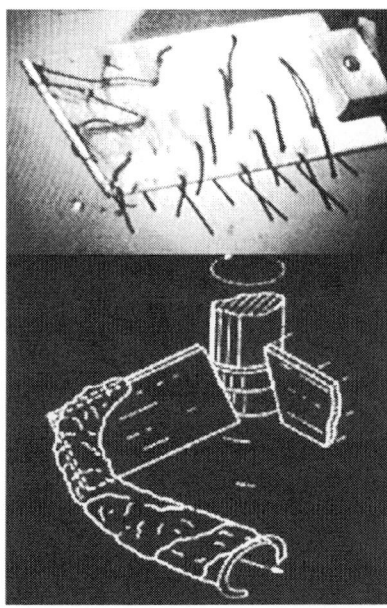

Figure 4: Pitched blade impeller vortex.

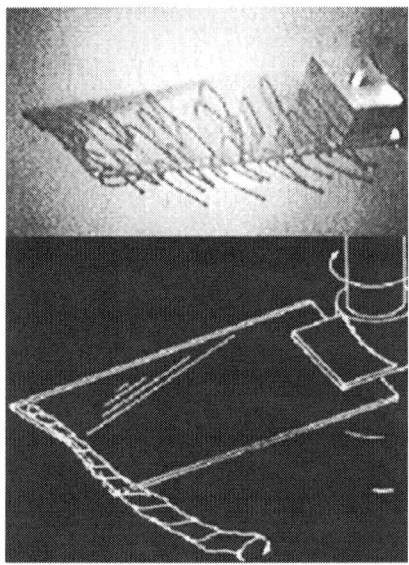

Figure 5: HE-3 impeller vortex.

In these figures, the impellers are rotating clockwise when viewed from above. Thus, the blades are moving into the page, and the view is of the backside of the blades. These vortices cause very localized wear emanating from the back-side or low-pressure side of impeller blades. These impeller vortices begin to diminish at Reynolds numbers below 10,000, become very weak below a Reynolds number of 500, and have completely disappeared below a Reynolds number of 10. As most slurry particles are heavier than the fluid, the centrifugal effect caused by the vortex will cause particles caught in the vortex to migrate to the OD of the vortex. Thus, the concentration of solids at the periphery of the vortex is much higher, and the rate of solid particle to surface impacts is much greater, increasing locally the rate of erosion. Of course, particles must be small enough to be captured in these vortices before this effect would be observed. On an industrial scale, however, the greatest majority of slurry applications would have particles sufficiently small to be captured by these vortices.

There are a number of relatively efficient wide blade impellers used in solids suspension service. Fasano and Reeder [18] compared the erosion rate between a Chemineer Maxflo W impeller, Figure 6, and a standard 45 four-bladed-pitched impeller (refer to Figure 4).

Figure 6: Chemineer Maxflo W impeller.

For the same level of solid suspension, these impellers utilize the same impeller diameter at the same speed. Therefore, velocities at the impeller are the same. The rate of erosion however for the pitched-blade impeller was, on a percentage basis, 59% greater than the erosion rate of the Maxflo W impeller.

Radial flow impellers are not very efficient in suspending solids. However, radial flow impellers are efficient in dispersing gasses. In applications where solids are present and a gas must be dispersed, they are often used. As with axial flow impellers, impeller efficiency changes with design. The Rushton or D-6 impeller, introduced in the late 1940s, creates a pair of vortices behind each blade in turbulent flow. These vortices, as we have observed with the pitched-blade impeller, can cause severe erosion from the backside of the blade. Figure 7 photo of the intertwined telltales on the backside of the blade demonstrates the size and nature of these vortices.

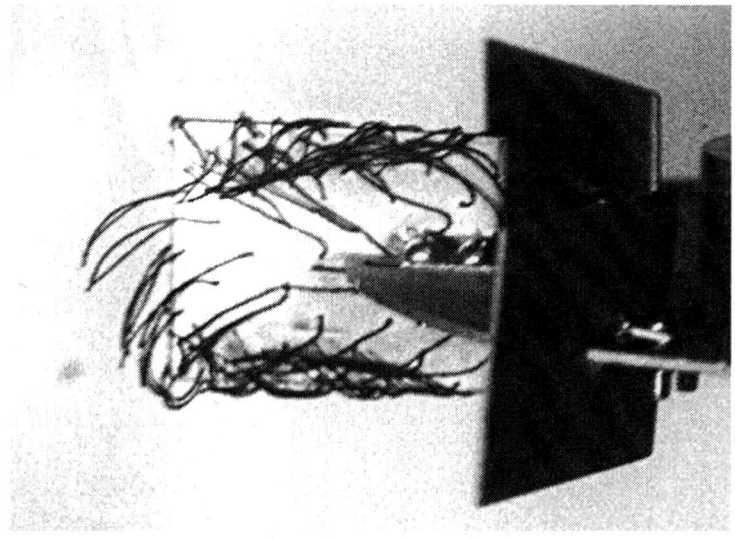

Figure 7: Telltales on D-6 impeller blade.

There exist today highly efficient radial flow impellers. The Chemineer CD-6 impeller was introduced into the marketplace in 1988, and the even more efficient Chemineer BT-6 was introduced in 1998. Both of these impellers exhibit very little tendency for vortex formation on the backside of the blade. This is demonstrated in telltale photos, in Figures 8 and 9. In each of these photos, the blades are rotating into the plane of the paper.

Figure 8: Telltales on CD-6 impeller blade.

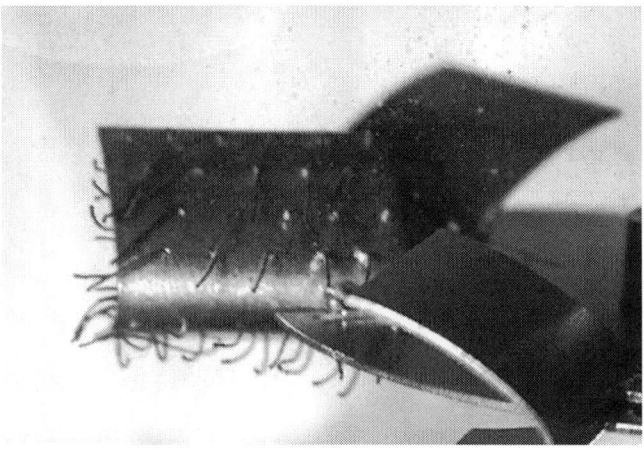

Figure 9: Telltales on BT-6 impeller blade.

Under turbulent flow conditions, it is usually very beneficial, both from an erosion standpoint as well as a process efficiency standpoint, to select a high-efficiency style impeller. In transitional flow, there are still benefits to be achieved by the use of a high-efficiency impeller, but not as profound as in turbulent flow. In laminar flow, there is no advantage in using a high-efficiency impeller either for erosion or solids suspension efficiency.

Horsepower and Speed Selection

As discussed earlier, erosion is very dependent on velocity, and typically in erosion-corrosion environments, the velocity exponent for the volumetric rate or weight rate of erosion is typically observed to be 2.5 to 4.0. In designing an agitator for the suspension of solids, the designer has a choice of selecting a number of power and speed combinations. Because solid suspension impeller efficiencies change with impeller style, impeller to tank diameter, and off-bottom clearance to tank diameter [19], a number of possible horsepower and speed choices can meet process objectives. The selection of a specific agitator design in the end should come down to economics. There are capital costs and operating costs. Capital costs are largely associated with the general size of the machine. The torque can generally best characterize the capital cost. Operating costs include the energy costs to operate the machine plus any maintenance costs. Maintenance costs include the costs of oil changes, new bearings, new gears, and new seal components, and in the case of erosion applications, the replacement of in-tank wear components, typically impeller blades. A close examination of the various horsepower and speed options should be examined closely in order to make an economic selection.

Example

For the sake of demonstration, let us assume that we need to design an agitator for suspending a 10% solution of sand in water. The sand will be assumed to have a weight mean particle size of 360 μm. The tank is 3.66 m diameter with a 2 : 1 elliptical dished bottom, and the water-sand slurry will have a liquid volume such that the depth of liquid in the tank is 3.66 m. The sand will be assumed to have a specific gravity of 2.4 and the water a specific gravity of 1.0. The viscosity of water will be assumed to be 1 mPa-s. The process solution requires that the solids be suspended to the "just suspended" condition such that no particles settle on the bottom of the tank for more than 2 seconds. All designs are to utilize a single HE-3 impeller one-third the tank diameter off-bottom. The horsepower, speed, and impeller diameter combinations that satisfy the process objective can be determined using solids suspension design procedures such as the article presented by Corpstein and others [19]. The following designs shown in Table 2 below all satisfy the off-bottom solids suspension process requirement.

Table 2: Possible process design selections for example

Impellerdiameter, m	Impellerdia./ tank dia.	Impellerspeed, rpm	Impellerpower, kW	Impellertorque, kNm	Impellertip speed, m/ min	Relativewear life*	Approx. relative cap. cost
0.889	0.243	159	3.49	0.210	444	1.00	1.00
1.016	0.278	123	2.94	0.228	392	1.54	1.06
1.143	0.313	102	3.15	0.294	366	1.96	1.25
1.220	0.347	92	3.59	0.372	367	1.95	1.45
1.397	0.382	84	4.05	0.460	369	1.92	1.67
1.524	0.417	75	4.61	0.587	359	2.10	1.95
1.651	0.451	70	5.23	0.713	363	2.02	2.22
1.778	0.486	67	6.32	0.972	374	1.82	2.71
1.905	0.521	65	9.16	1.345	389	1.59	3.35

*Relative wear based on assumed velocity exponent of 3.5.

Even though using small impellers operating at high speeds reduces capital costs and power, the high tip speeds of these designs lead to short wear lives. Using an intermediate size impeller that minimizes tip speed at the cost of higher capital and power costs maximizes the wear life. As can be observed, the torque and consequently the capital cost increase dramatically with impeller to tank diameter ratios greater than 0.45 due to changes in the flow pattern generated by the impeller.

Scale-Down Studies

Since there is often an erosion-corrosion synergistic effect that cannot be predicted a priori with today's current data, any material selection should be studied on a smaller scale, before making a final choice. Since it is important to model the same hydraulic behavior over the blade, the authors recommend that the ratio of mean particle diameter to impeller diameter not exceed 0.008. For example, a mean particle diameter of 1 mm would suggest a small-scale impeller no less than 125 mm (4.9 in) diameter. It is also important to ensure that fluid regimes have not changed. If the impeller operation is turbulent on the large scale, it should also be turbulent on the small scale.

An optimum agitator horsepower and speed selection can be made as described above for the full scale. However, in order to determine the rate of erosion, scale-down studies should be made. A geometric scale-down for an equal level of solid suspension will result in a tip speed that will always be lower on the smaller scale, except when scaling down geometrically for very slow settling solids (<0.1 m/min). Wear rate, as previously demonstrated, is a strong function of velocity. Therefore, all scale-down tests should be made at equal tip speed. As an example, if we examine the 1.397 m impeller solution for the above described problem and scale this down to a 0.4572 m diameter tank, we would have the comparison provided in Table 3.

Table 3: Comparison of equal suspension to equal tip speed

Condition	Impeller dia., cm	Impeller speed, rpm	Impeller tip speed, m/min
Full scale design, 144 in. dia. tank	139.7	84	369
Scale-down design based on equal suspension	17.46	456	250
Scale-down design based on equal tip speed	17.46	672	369

CONCLUSIONS AND SUMMARY

The rate of erosion is dependent on the following major static environmental factors: chemical environment, hardness of the solid particles, density of the solid particles, percent solids, the shape of the solids, the size of the solids including whether or not tramp solids are present, and type of impeller. The dynamic factors affecting erosion rate are fluid regime, impact velocity, impact frequency, and angle of impact. As there are no good means currently of predicting erosion rate, small-scale studies should be conducted emulating as much of the total environment as possible. These small-scale studies should be conducted using equal tip speed to mimic the full-scale rate of erosion.

Erosion in most mixing processes is a fatigue process normally accelerated by a liquid corrosive environment. The fatigue process occurs on a micro- or localized scale, and, as with macroscale fatigue, two stages of the erosion process have been observed. There is an incubation period followed by the formation and growth of pits involving the removal of the metal or material. One of two routes is generally utilized in dealing with an erosion application. The high-velocity areas such as the blades are either made from hard materials or coated with hard ceramic materials such as tungsten carbide or silicon carbide. Alternatively, the blades are covered with some type of elastomeric covering.

Impeller selection is important especially in turbulent flow conditions. High-efficiency impellers will generally erode at a slower rate because the backside of the blades has minimized shedding vortices. In laminar flow, from an erosion standpoint, most impellers behave similarly due to a lack of vortices. Thus, the selection of the impeller should be based primarily on what is required to accomplish the desired process result.

A number of horsepower and speed selections that satisfy the process requirement should be examined to conduct an economic analysis. The lowest possible speed may not be the most economical. It is best to first design the most cost optimum agitator for the full scale. Then in order to estimate the corrosion rate, scale down on the basis of equal tip speed.

REFERENCES

1.   R. Chattopadhyay, Surface Wear, Analysis, Treatment and Prevention, ASM International, Metals Park, Ohio, USA, 2001.

2.   S. G. Sapate and A. V. RamaRao, "Effect of erodent particle hardness on velocity exponent in erosion of steels and cast irons," Materials and Manufacturing Processes, vol. 18, no. 5, pp. 783–802, 2003.

3.   M. M. Stack, F. H. Stott, and G. C. Wood, "The significance of velocity exponents in identifying erosion-corrosion mechanisms," Journal de Physique IV, Colloque C9, Supplement au Journal de Physique III, vol. 3, 1993.

4.   A New Slurry Pump Standard, Pumps and Systems, The Hydraulic Institute, 2006.

5.   I. Fort, J. Medek, and F. Ambros, "Erosion wear of axial flow impellers in a solid-liquid suspension,"Acta Polytechnica, vol. 41, no. 1, pp. 23–28, 2001.

6.   Y. Zheng, Z. Yao, X. Wei, and W. Ke, "The synergistic effect between erosion and corrosion in acidic slurry medium," Wear, vol. 186-187, no. 2, pp. 555–561, 1995.

7.   V. N. Amelyushkin and B. N. Agafonov, "Special features of erosion wear of rotor blades of cogeneration steam turbines," Power Technology and Engineering, vol. 36, no. 6, pp. 359–362, 2002.

8.   Y. A. Khalid and S. M. Sapuan, "Wear analysis of centrifugal slurry pump impellers," Industrial Lubrication and Tribology, vol. 59, no. 1, pp. 18–28, 2007.

9.   D. López, J. P. Congote, J. R. Cano, A. Toro, and A. P. Tschiptschin, "Effect of particle velocity and impact angle on the corrosion-erosion of AISI 304 and AISI 420 stainless steels," Wear, vol. 259, no. 1-6, pp. 118–124, 2005.

10. R. C. Corpstein and J. B. Fasano, "Erosion of rubber covered impeller blades in an abrasive service,"The Indian Chemical Engineer, vol. 36, no. 1, 1990.

11. J. Wu, B. Ngyuen, L. Graham, Y. Zhu, T. Kilpatrick, and J. Davis, "Minimizing impeller slurry wear through multilayer paint modelling," Canadian Journal of Chemical Engineering, vol. 83, no. 5, pp. 835–842, 2005.

12. J. E. Miller and F. Schmidt, Slurry Erosion: Uses, Applications and Test Methods, ASTM, 1987.

13. G. W. Stachowiak and A. W. Batchelor, Engineering Tribology, Butterworth-Heinemann, 3rd edition, 2005.

14. R. W. Armstrong and F. J. Zerilli, "Dislocation mechanics based viscoplasticity description of FCC, BCC and HCP metal deformation and fracturing behaviors," in Proceedings of ASME International Mechanical Congress and Exposition, pp. 417–428, November 1995.

15. K. C. Wilson, G. R. Addie, A. Sellgren, and R. Clift, Slurry Transport Using Centrifugal Pumps, Springer, 2nd edition, 2005.

16. D. S. Dickey and J. B. Fasano, Handbook of Industrial Mixing, chapter 21, section 9, John Wiley and Sons, New Jersey, NJ, USA, 2004, Edited By Paul, Atiemo-Obeng and Kresta.

17. J. B. Fasano, "Flow visualization techniques on rotating impellers," in Proceedings of the Engineering Foundation Mixing Conference XII, Pitosi, MO, USA, 1989.

18. J. B. Fasano and M. F. Reeder, "An improved maxflo impeller," in Proceedings of the North American Mixing Forum, Mixing Conference XVII, Banff, Canada, August 1999, Paper 2.4.

19. K. J. Myers, R. R. Corpstein, A. Bakker, and J. B. Fasano, "Solid suspension agitator design with pitched blade and high efficiency impellers," in Proceedings of the AIChE Annual Meeting, St. Louis, MO, USA, November 1993.

Chapter 9

Design of a Turbulence Generator of Medium Consistency Pulp Pumps

Hong Li, Haifei Zhuang, and Weihao Geng

Research Center of Fluid Machinery Engineering and Technology of Jiangsu University, Zhenjiang, Jiangsu 212013, China

ABSTRACT

The turbulence generator is a key component of medium consistency centrifugal pulp pumps, with functions to fluidize the medium consistency pulp and to separate gas from the liquid. Structure sizes of the generator affect the hydraulic performance. The radius and the blade laying angle are two important structural sizes of a turbulence generator. Starting with the research on the flow inside and shearing characteristics of the MC pulp, a simple mathematical model at the flow section of the shearing chamber is built, and the formula and procedure to calculate the radius of the turbulence generator are established. The blade laying angle is referenced from the turbine agitator which has the similar shape with the turbulence generator, and the CFD simulation is applied to study the different flow fields with different blade laying angles. Then the recommended blade laying angle of the turbulence generator is formed to be between 60° and 75°.

INTRODUCTION

MC (Medium Consistency) pulp pumps are key equipment to transport pulp in modern paper mills. Paper pulp suspension contains three kinds of media, namely, fibers, water, and air, which results in a high flow complexity and particularity in the pump itself. The pulp cannot move when the pulp mass consistency is more than 6%. In order to transport the medium with the consistency over 6%, the MC pulp pumps must have the ability to fluidize the pulp. The turbulence generator is the key component of the MC centrifugal pulp pump, which fluidizes the MC pulp and separates the gas from the pulp suspension. The structure of MC pulp pumps is shown in Figure 1.

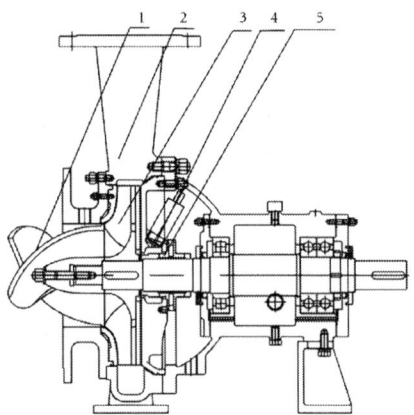

Figure 1: MC pulp pump. (1) Turbulence generator. (2) Casing. (3) Impeller. (4) Gas outlet. (5) Mechanical seal.

Some research had been done in China [1–4], including the simulation of fluidization of paper pulp suspension and the optimized designs of the turbulence generator. But the systemic design theories and methods are still not built. Starting with the research on the flow and shearing characteristics of the MC pulp, a simple mathematic model is built, and the formula for calculating the radius of the turbulence generator is established. The range of the blade laying angle is obtained by the recommended blade laying angle from the turbine agitator which has the same shape as the turbulence generator, and the CFD simulation is applied to study the different flow fields with different blade laying angles.

DEDUCING THE CRITICAL SHEAR VELOCITY GRADIENT

The flow of MC pulp suspension is neither similar to usual water flow, nor to two-phase flow nor liquid-particle flow. It is a kind of three-phase fluid, consisting of gas-liquid (water)-solid (fiber) flow. It has very complicated flow characteristics, changing with the species, consistency, and velocity of the paper pulp and the fiber shape.

High consistency of the fibers and gas/air in the MC pulp makes the fiber suspensions fail to move forward freely. By high-speed rotation, the turbulence generator introduces high shearing force to distribute fibrous reticulum and also avoid fibers to flocculate again. In this situation, pulp fiber suspensions show the flow characteristics as similar to water. Therefore it is defined as fluidization [4].

The minimum shear force which makes fibrous reticulum of the MC pulp suspension distributed is called the critical shear force. Based on experimental researches on sulphated wood pulp by Hemstrom et al. [5], the critical shear force τ_d can be given as

$$\tau_d = KC^\alpha,$$

(1)

where k and a are the coefficients related to the species of the paper pulp. For given species of the paper pulp, the critical shear force τ_d is only related to C, which is the mass consistency of the paper pulp suspension.

The apparent viscosity of the non-Newtonian fluid is defined as the ratio of the shear force to the shear velocity [6]. The apparent viscosity, μ_a (Pa·s), is given by

$$\tau_d = \mu_a \frac{du}{dr} = \mu_a S_w,$$

(2)

where τ_d is the shear force (Pa) and S_w is the shear velocity gradient (1/s). We can get the critical shear velocity gradient of the paper pulp by (1), but for the actual generator design, there is no direct relationship between the critical shear force and the geometric parameters. So we need to change the condition parameter of the fluidization from the critical shear force to the critical shear velocity gradient, by analyzing the relationship between the shear force, the shear velocity, and the shear velocity gradient.

Duffy et al. [7] obtained the apparent viscosity of straw wood pulp by experiments. The apparent viscosity, μ_a, is given by

$$\mu_a = 0.178 C^{3.30} S_w^{-0.75},$$

(3)

simultaneously by (2) and (3)

$$\tau_d = \mu_a \frac{du}{dr}$$

$$= 0.178 C^{3.30} S_{wd}^{-0.75} \cdot S_{wd}$$

$$= 0.178 C^{3.30} S_{wd}^{0.25},$$

(4)

where S_{wd} is the shear velocity gradient for the fluidization of paper pulp.

Simultaneously by (1) and (4)

$$KC^{\alpha} = 0.178 C^{3.30} S_{wd}^{0.25},$$

$$S_{wd} = \left(5.618 K C^{\alpha-3.30}\right)^4.$$

(5)

Gullichsen and Harkonen [8] and Kefu [9] obtained the values of K and α with shearing experiments on MC pulp. Based on the experimental data and the formulations above, K, α, τ_d (C = 15%) and S_{wd} of seven common kinds of paper pulp are obtained, as shown in Table 1.

Table 1: K, α, τ_d (C = 15%) and of seven kinds of paper pulp

Species of paper pulp		α	τ_d	S_{wd}
Bleaching poplar wood pulp	27.3	1.98	5818.66	341.4
Unbleached poplar wood pulp	18.9	2.04	4739.01	150.2
Spruce wood pulp	6.7	2.43	4830.33	162.1
Bleaching redpine sulphate wood pulp	5.74	2.52	5286.34	231.5
Unbleached redpine sulphate wood pulp	5.38	2.52	4949.17	216.9
Unbleached stone ground wood pulp	0.4	3.49	5088.84	199.7
Unbleached waste paper pulp	0.15	3.8	4073.4	113.5

PARAMETERS OF THE TURBULENCE GENERATOR

The turbulence generator in this research consists of a hub and three blades. The blade working face is perpendicular to the surface of the hub. Outside surfaces and inside surfaces of blades are all cylindrical surfaces.

Main parameters of a turbulence generator are shown in Figure 2.

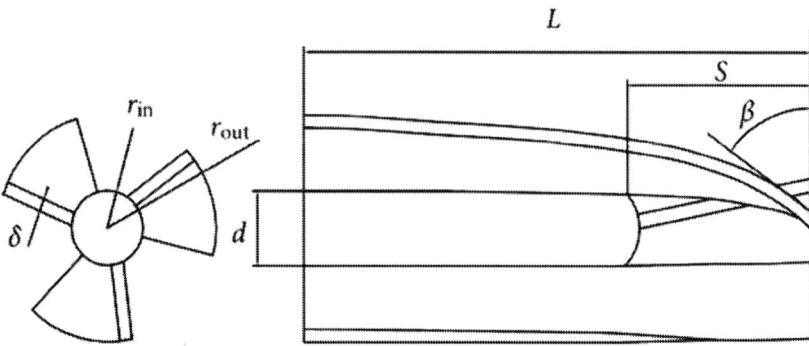

Figure 2: Turbulence generator. r_{out} is the blade outlet radius; r_{in} is the blade inlet radius; d is the hub diameter; δ is the blade thickness; β is the blade laying angle; S is the blade overhang length; L is the blade total length.

The blade outlet radius and the blade laying angle are the most important design parameters, which are the preconditions of other structure sizes, deciding the working range and efficiency of the MC pulp pumps.

BLADE OUTLET RADIUS OF TURBULENCE GENERATOR

In this part, starting with the researches on the flow characteristics of the MC pulp suspension, a series of formulas are deduced to qualify the critical fluidization of MC pulp. The formulas for calculating the radius of the turbulence generator have then been established by simplifying the model of the flow field of the turbulence generator.

Calculating Blade Outlet Radius

As shown in Figure 3, supposing that the paper pulp suspension fills within the full flow passage, we analyze the force of laminar flow in the torus field at a section of the shearing chamber of the turbulence generator.

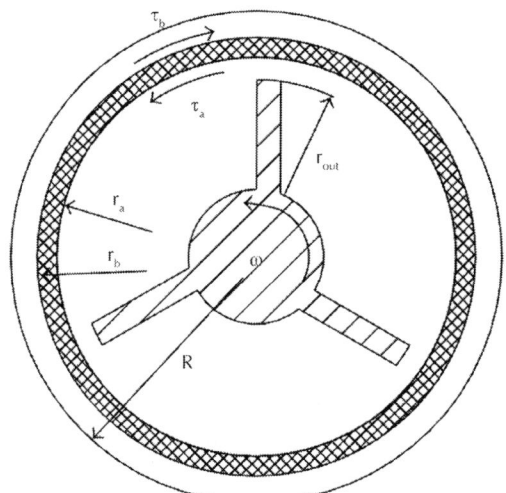

Figure 3: Flow at a section of the shearing chamber of the turbulence generator.

When the turbulence generator is rotating at a high speed, neglecting axial velocity, we can consider the paper pulp flow being laminar in the area from blade top of the turbulence generator to the inside wall of the shearing chamber. We note the shear force inside as τ_a and outside as τ_b, as shown in Figure 3. At the normal running condition, the momentum moment to axis in the grid flow field keeps constant. So the whole moment is zero. That is,

$$2\pi r_a \cdot \tau_a \cdot r_a = 2\pi r_b \cdot \tau_b \cdot r_b.$$

(6)

So,

$$\tau(r) \cdot r^2 = M,$$

$$\tau(r) = \frac{M}{r^2},$$

(7)

where M is a constant and $\tau(r)$ is the circumferential shear force at r radius position.

Based on formulas above, given the shearing chamber radius R and the shear force at the inside wall $\tau(R)$, we can obtain the momentum moment M:

$$M = \tau(R) \cdot R^2.$$

(8)

Combine (4), (7), and (8). S_w is given by,

$$0.178C^{3.30}S_w(r)^{0.25} = \frac{\tau(R) \cdot R^2}{r^2},$$

$$S_w(r) = \left(\frac{\tau(R) \cdot R^2}{0.178C^{3.30}r^2}\right)^4,$$

(9)

where $S_w(r)$ is the radial velocity gradient when the radius is r.

In conditions of a laminar flow, the shear velocity gradient is inversely proportional to the distance to the rotating axis in the area from the blade top of the turbulence generator to the inside wall of the shearing chamber. So the shear force near the inside wall of the shearing chamber is the smallest. We can consider that the whole flow field becomes turbulent, if the shear velocity gradient near the inside wall reaches the critical value.

Define N as

$$N = \left(\frac{\tau(R) \cdot R^2}{0.178C^{3.30}}\right)^4.$$

(10)

where N is a constant which is decided by the species of the paper pulp and the shearing chamber radius R.

So (9) becomes,

$$S_w(r) = N\frac{1}{r^8}.$$

(11)

The velocity of pulp at the blade outlet of the turbulent generator is given by

$$v_{out} = \int_{r_{out}}^{R} s_w(r)dr$$

$$= N\left(\frac{1}{7r_{out}^7} - \frac{1}{7R^7}\right),$$

(12)

$$v_{out} = 2\pi r_{out}\frac{n}{60}.$$

(13)

Combining (12) and (13) gives

$$r_{out} = \frac{30N}{7\pi n}\left(\frac{1}{r_{out}^7} - \frac{1}{R^7}\right).$$

(14)

From the formulas above, we can get the critical shear force τ_d from Table 1. Let $\tau(R)$ be equal to τ_d, and we get the minimal from (10). Then r_{out} can be obtained from (14).

Design Example

The design parameters of an MC pulp pump are taken as follows: Q = 60 m³/h, H = 50 m, n=1450 r/min, C = 8%–15%, and the pump inlet diameter (shearing chamber diameter) D = 150 mm.

Based on the parameters above and Table 1 when C is 15%, the critical shear force τ_d = $\tau(R)$ = 5818.66 Pa, the shearing chamber radius R = D/2 = 75 mm. So we get the following from (13):

$$N = \left(\frac{5818.66 \times 0.075^2}{0.178 \times 15^{3.30}}\right)^4 = 3.417 \times 10^{-7}.$$

(15)

Finally, we obtain r_{out} from (14) that $r_{out} = 62.47$ mm. Let $r_{out} = 65$ mm after roundness.

By the experiment of the MC pulp pump with the turbulence generator of the sizes given above, the performance of the pump is checked and achieved in the paper published before [10]. This paper introduced the main structure of the test bed for the centrifugal pulp pump. The test result showed that the pump could run stably and efficiently under the 11% stock consistency in that test condition, with the efficiency to 40%. It testified the MC pump had a good performance, satisfying the pulp transporting needs.

BLADE LAYING ANGLES OF TURBU-LENCE GENERATOR

Setting Blade Laying Angle

The principle of the turbulence generator is similar to an agitator. The blade structure is similar to a pitched turbine agitator, as shown in Figure 4. The viscosity coefficient of the transporting medium in one agitator can reach 100 Pa.s, which matches up to that in the turbulence generator of the MC pulp pump. According to [11], the recommended blade laying angles are 45°, 60°, and 90°. The initial axial velocity inside the agitator is usually zero, but it is not null inside the pump. Compared to the recommended blade laying angle of the pitched turbine type agitator, the angle of the turbulence generator should be modified according to the actual flow.

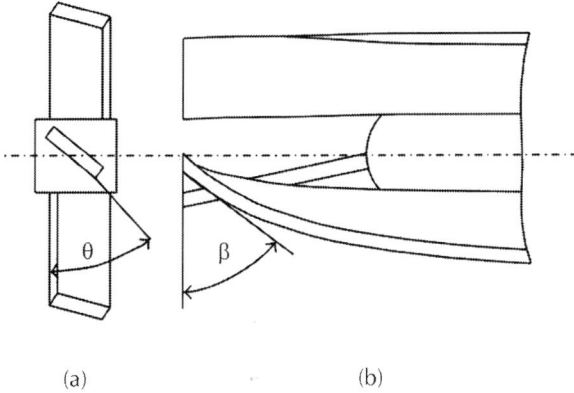

(a)   (b)

Figure 4: Pitched turbine type agitator (a) and turbulence generator (b).

Figure 5 shows the outlet streamline of the turbulence generator. α is the correction angle, which means the angle between relative velocity w direction and circumferential velocity direction at the blade outlet; θ is the recommended blade laying angle of the turbine agitator; β is the blade laying angle of outlet streamline.

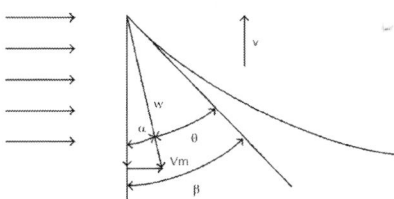

Figure 5: Velocities at the outlet streamline of the turbulence generator.

We can obtain the following equations from Figure 5

$$\beta = \alpha + \theta,$$

(16)

$$\alpha = \arctan \frac{v_m}{v},$$

(17)

$$v_m = \frac{Q}{\pi R^2},$$

$$(18)$$

where Q is the volumetric flow (m³/h); v_m is the axial velocity (m/s); w is the relative velocity (m/s); v is the linear velocity at the blade outlet (m/s); R is the shearing chamber radius of the turbulence generator (m).

In the application of the MC pulp pump, the rotating direction of the turbulence generator should be contrary to the pump impeller. The blade laying angle cannot exceed 90°, so α should be 0° when θ is already 90°.

The blade correction angle α is 10.8° after calculation by (17). According to (16), three blade laying angles β of the turbulence generator are 55.8°, 70.8°, and 90°, respectively.

CFD Simulation Model

CFD (Computational Fluid Dynamics) is used to calculate and analyze the flow field through solving basic equations, such as the momentum conservation equations, the mass conservation equations, and the energy conservation equations. The numerical simulations in this paper are performed using FLUENT 6.2.

The flow of the MC pulp suspension after fluidization is a turbulence flow, and the flow characteristics are similar to the gas-water two-phase flow, but the large amount of gas inside the pulp will influence the flow.

The research on the inner flow in the MC pulp pump focuses on the movement and distribution of gas in the paper pulp, interaction between gas and liquid, as well as the turbulence distribution in the flow field. So in the CFD simulation of the fiber suspension flow in the shearing chamber, we apply Eulerian gas-liquid phase model and RNG - model, with RANS as momentum equations [12]. We use the MRF (Multiple Reference Frames) [13, 14] to build the frame, Phase Coupled SIMPLE to deal with the coupling of pressure and velocity, and the implicit steady-state segregating solution to solve the equations.

Building 3D Models and Boundary Conditions

We establish a 3D model and define the grid for the flow field of the turbulence generator when the blade laying angles are 55.8°, 70.8° and 90°, respectively. We make the field a whole field to simplify the computation, as shown in Figure 6. Boundary conditions are determined as the real running conditions of the turbulence generator. The gas volume fraction is 20%.

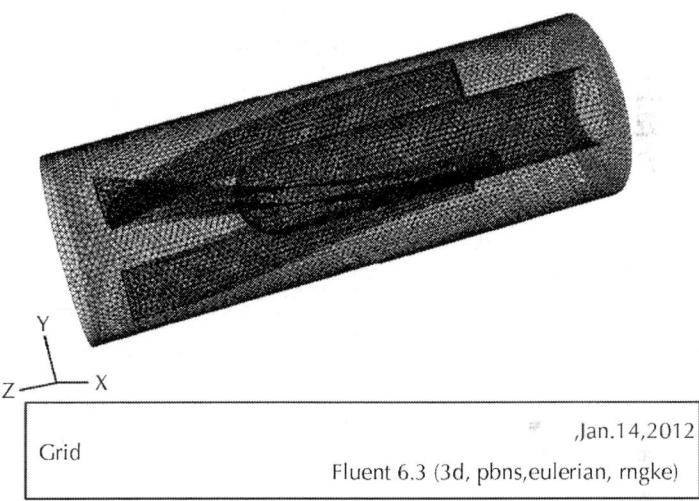

Grid

,Jan.14,2012

Fluent 6.3 (3d, pbns,eulerian, rngke)

Figure 6: The grid of a turbulence generator.

Analysis of the Simulation Results

Simulation results are compared and analyzed, which are the effects on the gas-liquid separation, the turbulent kinetic energy, the torque produced by the turbulence generator, and the pressure changing from the outlet to the inlet of the turbulence generator.

Effects on Gas-Liquid Separation

From Figure 7, we can observe that there is no big difference of effects on gas-liquid separation, with the gas volume fraction reaching almost

95% at the end of the turbulence generator. The blade impact angle from the medium will increase with the enlargement of the blade laying angle, which results in the low pressure field being strengthened and expanded on the suction face at the blade front part. So we can find that the bigger the blade laying angle is, the shorter the gas-liquid separation field is to the blade front part.

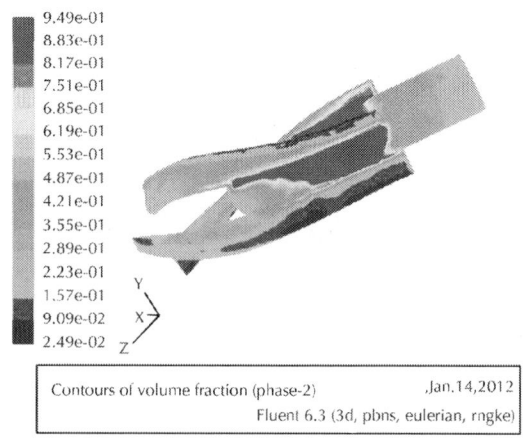

(a)

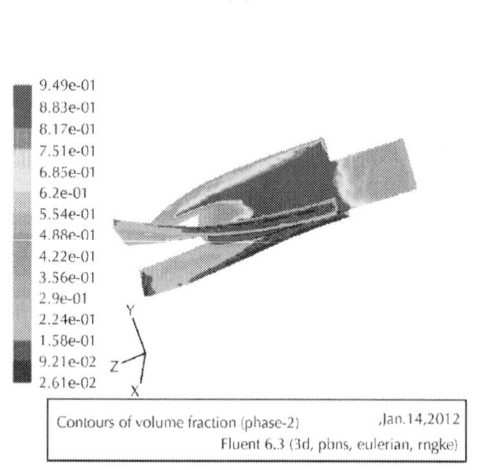

(b)

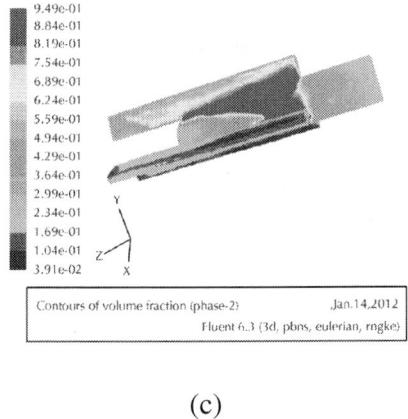

(c)

Figure 7: Gas distribution on the blade surfaces with different blade laying angles.

Turbulent Kinetic Energy Distribution

According to Figure 8, the turbulent kinetic energy in the field between the blade outlets to the inside wall of the shearing chamber is lower than that around blades. And the distribution of the turbulent kinetic energy trends to be well distributed with the increase of the blade laying angle.

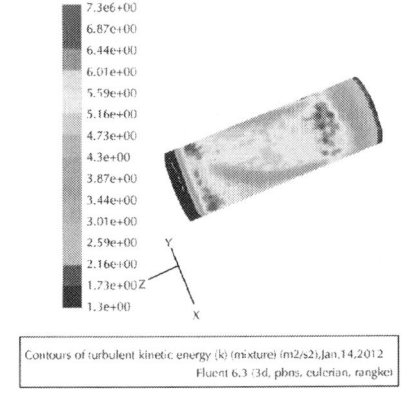

(a)

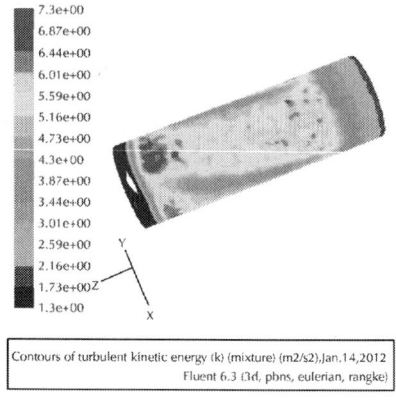

(b)

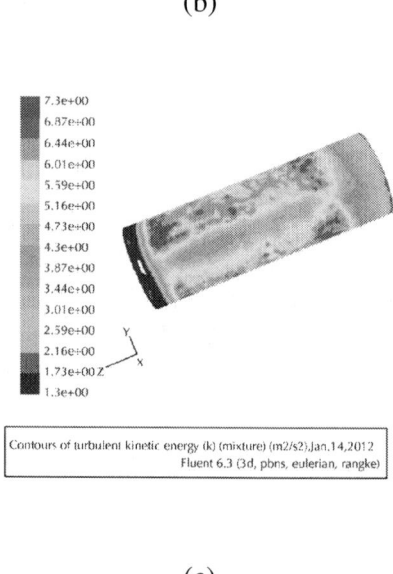

(c)

Figure 8: Turbulent kinetic energy distribution in shearing chamber with different blade laying angles.

Targeted Values

From Table 2, we can get that, with the increase of the blade laying angles β, the turbulence intensity will also increase, so did the torque value of the turbulence generator. But the differential pressure in the flow field from outlet to inlet decreases continuously. Theoretical analyses show that, with the increase of the blade laying angles β, the guiding role that blades were playing for the paper pulp decreases, which causes the torque value produced by the turbulence generator to be increased. Meanwhile the increase of the impact from the blades to the medium makes the flow more complicated, which results in a stronger turbulence flow. When blade laying angle β becomes small, the axial thrust applied to the paper pulp produced by the blades makes the outlet pressure bigger than the inlet pressure. When β increases to a certain value, the axial trust will offset the frictional head loss as of the paper pulp flow. There will be no axial thrust when β is 90°, and the frictional loss causes the outlet pressure to become smaller than the inlet pressure. Based on the results of the numerical simulation, the recommended blade laying angle of the turbulence generator is from 60° to 75°.

Table 2: Targeted values with different blade laying angles

	Turbulent kinetic energy K (m²/s²)	Torque value (N m)	Differential pressure (Pa)
55.8°	131.28%	24.56	5721.83
70.8°	135.72%	25.34	1676.43
90°	140.39%	26.58	−6404.76

CONCLUSIONS

- We analyzed the principle and the conditions of the fluidization of medium consistency pulp. We established the fluidization expression formulas with the characteristic parameters as variables of the MC pulp pump.
- We established the flow mathematical model inside the shearing chamber and obtained the formula to calculate the blade outlet radius of the turbulence generator.
- We simulated the pulp flow inside the turbulence generator. Blade laying angle of the turbulence generator has a small influence on the gas-liquid separation. But the bigger the blade laying angle

is, the nearer the gas-liquid separation field is to the blade front part. And the distribution of the turbulent kinetic energy trends to be well distributed with the increase of the blade laying angles.

- The recommended blade laying angle of the turbulence generator is from 60° to 75°.

ACKNOWLEDGMENT

This paper is is supported by high technology project of Jiangsu province, China, ID: SBE201000567, R&D of MC paper pulp pumping system.

REFERENCES

1.   Z. Jiankang, X. Zonghua, C. Kefu, et al., "Design of turbulence medium consistency pulp pump-ZBJ31,"China Pulp and Paper, vol. 3, no. 6, pp. 3–8, 1986.

2.   Q. F. Chen, K. F. Chen, and R. D. Yang, "Study on the structure of turbulent generator of MC centrifugal pump based on CFD," China Pulp and Paper, vol. 25, no. 10, pp. 25–27, 2006. ·

3.   L. Hong, Z. Wanyi, Z. Rongsheng, et al., "Structure characteristics of centrifugal pulp pump," China Pulp and Paper, vol. 22, no. 4, pp. 38–40, 2003.

4.   C. Qifeng, C. Kefu, Y. Rendang, et al., "Study on the fluidized characteristics of medium-consistency pulp suspensions," Journal of China Pulp and Paper, vol. 18, no. 12, pp. 148–150, 2003.

5.   G. Hemstrom, K. Moller, and B. Norman, "Boundary layer studies in pulp suspension flow," Tappi, vol. 59, no. 7, pp. 115–118, 1976. ·

6.   S. Wence, Engineering Fluid Mechanics, Dalian University of Technology Press, Dalian, China, 2004.

7.   G. G. Duffy, K. Moller, P. F. W. Lee, et al., "Design correlations for groundwood pulps and the effects of minor variables on pulp suspension flow," Appita, vol. 27, no. 5, p. 327, 1974.

8.   J. Gullichsen and E. Harkonen, "Medium consistency technology—1. Fundamental data.," Tappi, vol. 64, no. 6, pp. 69–72, 1981. ·

9.   C. Kefu, Technology and Equipment of Medium-High-Consistency Pulp, South China University of Technology Press, Guangzhou, China, 1994.

10. L. Hong, G. Gweihao, F. Jianguo, et al., "Experimental study on medium consistency centrifugal pulp pump," China Pulp and Paper, vol. 27, no. 2, pp. 71–72, 2008.

11. C. Zhiping, Z. Xuwen, and L. Xinghua, Mixing and Mixing Equipment Application Manual, Chemical Industry Press, Beijing, China, 2004.

12. C. Qifeng, Study on medium-consistency pulp suspension and CFD application, Ph.D. dissertation, South China University of Technology, Guangzhou, China, 2005.

13. G. Z. Zhou, L. T. Shi, and Y. C. Wang, "Computational fluid dynamics progress in stirred tank reactors," Chemical Engineering (China), vol. 32, no. 3, p. 28, 2004. ·

14. H. Sun, W. Wang, and Z. Mao, "Numerical simulation of whole three-dimensional flow field in stirred tank with anisotropic turbulence model," Journal of Chemical Industry and Engineering (China), vol. 53, no. 11, pp. 1153–1159, 2002. ·

Chapter

10

Modeling Optimal Scheduling for Pumping System to Minimize Operation Cost and Enhance Operation Reliability

Yin Luo, Shouqi Yuan, Yue Tang, Jianping Yuan,
and Jinfeng Zhang

Research Center of Fluid Machinery Engineering and Technology,
Jiangsu University, Zhenjiang 212013, China

ABSTRACT

Traditional pump scheduling models neglect the operation reliability which directly relates with the unscheduled maintenance cost and the wear cost during the operation. Just for this, based on the assumption that the vibration directly relates with the operation reliability and the degree of wear, it could express the operation reliability as the normalization of the vibration level. The characteristic of the vibration with the operation point was studied, it could be concluded that idealized flow versus vibration plot should be a distinct bathtub shape.

There is a narrow sweet spot (80 to 100 percent BEP) to obtain low vibration levels in this shape, and the vibration also follows similar law with the square of the rotation speed without resonance phenomena. Then, the operation reliability could be modeled as the function of the capacity and rotation speed of the pump and add this function to the traditional model to form the new. And contrast with the tradition method, the result shown that the new model could fix the result produced by the traditional, make the pump operate in low vibration, then the operation reliability could increase and the maintenance cost could decrease.

INTRODUCTION

As important aspects in engineering industries, low cost and high reliability are the focus of the operation control in pumping system [1, 2].

The purpose of pump scheduling function is to schedule the operation of N pumps over a time period to meet consumer demands, and optimizing this function has been proven to be a practical and highly effective method in reducing operation costs without altering the actual infrastructure of the whole system. Thus, this issue naturally draws the attention of researchers [2].

Pump system scheduling should be robust with any operation scenarios and should deal virtually with all operation factors, such as variable speeds, constant speeds, and switched-off pumps, in relation with operation constraints relative to power, head, flow, and speed. It is a very complex problem.

Many researchers have developed optimal control concepts to minimize operating costs associated with water-supply pumping systems. Mays [3] listed and classified various algorithms that have been developed to solve the associated control problem. In earlier studies, linear, nonlinear, integer, dynamic, mixed, and other kinds of programming were used to optimize a single objective: reduction of the electric energy cost. A detailed review of these works can also be found in [4]. Later, Lansey et al. [5] introduced the number of pump switches as an alternative to evaluate the pumps' maintenance costs, which became the second objective considered until that time. This method also proved that one Gulf Coast refinery project [6] can save about $2 million per year. In this way, as far as pump operation is concerned, the basic optimal scheduling model to reduce electrical and maintenance costs was formulated. The development of the pump scheduling model was later modified by forming or adding some optimization objectives, such as reservoir level variation and power peak, according to specific conditions.

However, in contrast to the estimation of electrical cost, where the computation is straightforward, computing for maintenance costs using the number of pump switches has some limitations.

According to some researchers [7–10], the cost of unscheduled maintenance which was not considered in former model may be the largest contributor to operation cost in process plants, and although with the mainly aspect of the wear and tear in switch course, the operation course should not been ignored.

At the same time, Bloch and Geitner [7] published three hydraulic factors, namely, rotation speed, impeller diameter (tip clearance), and operation point that can affect operation reliability which affects the unscheduled maintenance directly.

Subsequently, further improvements can be realized if these hydraulic factors could be quantified to develop an objective model in pump scheduling to enable the pump to operate in a high-reliability mode, then the reliability incidents would be reduced. As a result, maintenance cost can be reduced [10].

Vibration is one of the most vexing problems of pumping machineries, and it is the primary cause of considerable altercations and litigations. Excessive vibration of pumps and piping can destroy parts of the equipment (such as drive shafts, bearings, and seals). It can affect the reliability and life of the equipment and is often assumed as direct sign of reliability and health of the machine. One major end user puts a great deal of emphasis in reducing pump vibration through precise maintenance program, and some researchers adopted this index to indicate the degree of health and reliability of pumps. For example, Perez proposed a new method based on vibration called Nelson plot to assess the risk of low-flow operation in centrifugal pumps [11].

The main purpose of the current study is to set the vibration level as quantification of the hydraulic factors that influence operation reliability and the wear and tear in operation. Based on analysis of the vibration characteristics, an objective model is formulated to describe the relative reliability during operation, which presents a quantitative approach to evaluate alternative operating conditions. This new objective is added to the traditional model to form a new scheduling model. Subsequently, the operating conditions of the pump can be improved by making the pump operate at low vibration, the maintenance cost will be reduced, and the operation reliability will be enhanced to a certain degree.

VIBRATION CHARACTERISTICS OF PUMPS

For centrifugal pumps, the sources or causes of vibration are mainly composed of two types: mechanical and hydraulic.

For the mechanical cause, some imbalance and shaft misalignment always exist. This can cause vibration of the pump, and the intensity of this kind of vibration is related to the excitation force caused by the imbalance of the load. The load is directly related to the operation point.

Hydraulic vibration is caused by the reaction of the impeller vane as it passes the casing cutwater. Pump operation offers the best efficiency point (BEP) and thus creates eddies within the pump and some flow instabilities, such as cavitation. All these factors directly correlate with the operation point.

As throttle and speed controls are the most common control methods in centrifugal pumping system, the vibration characteristics under these two control operating modes were analyzed.

Vibration Characteristics of the Pump under Throttle Control

Basically, three types of vibrations should be distinguished: free vibrations, forced vibrations, and self-excited vibrations. Whereas free vibrations are rarely significant in pump operations, forced and self-excited vibrations frequently cause problems. For the forced vibrations, the force is the cause of the vibration, and self-excited vibration occurs when the exciting frequency is close to the natural frequency; it is an abnormal condition that should be avoided in system design and installation. Therefore, the vibration characteristics can be obtained by studying the mechanical and hydraulic force characteristics.

A force curve is shown in Figure 1 [12], summarized by Vorhoeven through test results from 47 pumps. The horizontal axis represents the BEP multiples, and the vertical axis represents the dimensionless number of the forces, expressed as

$$K_R = \frac{F}{\rho g H D_2 b_2},$$

(2.1)

where F is the force, ρ is the fluid density, H is the pump head, D_2 is the outer diameter of

the impeller, and b_2 is the outlet width of the blade.

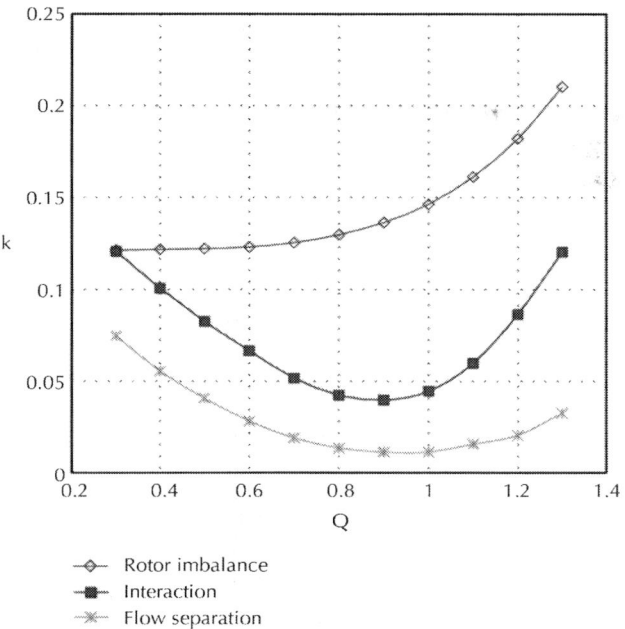

Figure 1: Forces in the impeller.

From the curve, the force caused by imbalance has a remarkable characteristic: it does not become small because of the BEP, and it is related to the load. When the load exceeds the BEP, the force significantly increases, and this force is relatively stable and small when the pump operates with partial load.

For the force generated by the flow, some features can be found. The minimum excitation is in the range of 0.9 BEP to BEP, and the excitation increases with partial load and at high flow condition. These curves are displayed in distinct bathtub shape. The stronger excitation caused by the interaction has a large slope at the right of the BEP, whereas the other has smaller slope at the left of the BEP.

From the above discussion, conclusion can be drawn that the idealized flow versus vibration plot should have a distinctive bathtub shape. For this bathtub shape, a narrow sweet spot (80% to 100% BEP) exists that can be used to obtain low vibration levels. The capacities move to the left and right of the low vibration level point, the vibration levels begin to increase, and the slope at the right area is larger than that at the left.

This bathtub-shaped flow versus vibration plot also can be found in some research works [9–11].

Figure 2 shows the test result of a well-designed centrifugal pump obtained by Ni et al. [13]. The pump was precision manufactured and installed. In Figure 2, BX, BY, and BZ are three direction vibration levels of the bearing house, whereas pc is the vibration level of the pump cover. The flow versus vibration plot in that test is exactly as the expression, and similar result can also be found in Perez's work [11].

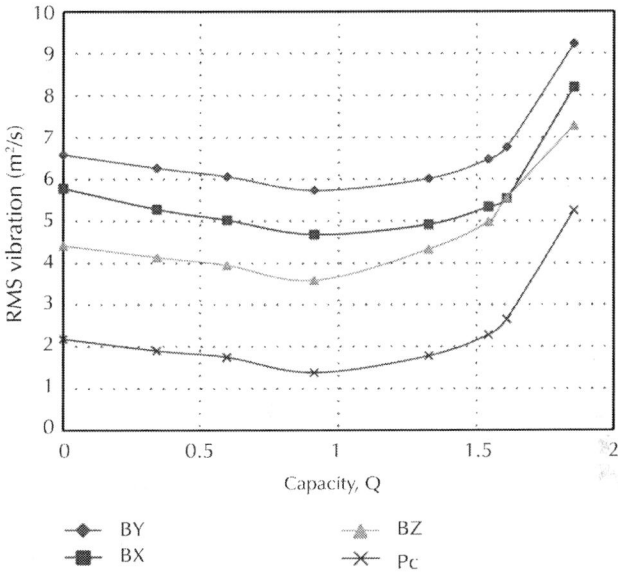

Figure 2: Vibration level in a well-designed centrifugal pump by Ni.

Vibration Characteristics of the Pump under Speed Control

Mechanical and hydraulic excitations are the main causes of vibration.

According to the affinity law in pumps, when a pump is operating at two different speeds, the flow condition in the pump is homologous. Thus, when the pump operates at different rotation speeds, the variable tendency of hydraulic excitation versus the flow is still the same as that at different speeds, but the degree of some flow phenomena, such as instability interaction, will change under different speeds, especially at low speed.

The mechanical excitation is mostly related to the balance degree and the load. The load still has some homologous regularities expressed as the variable tendency of hydraulic excitation versus flow, which is almost the same at different speeds. In addition, once the speed drops below the nominal motor speed, the mechanical excitation is reduced quite obviously.

Based on the above results, the excitation versus flow at different rotor speeds is noted to have similar varying tendency to a certain extent, but the degree of excitation is different. Therefore, the vibrations among different rotor speeds from this point have similar varying tendency to a certain extent.

Figure 3 [10] shows the test vibration level dates of a pump in a chemical process with model 1.5 × 3–13 ANSI B 73.1. The measured point of vibration is at the thrust bearing horizontal plane, and the date was adopted with root mean square (RMS) expression. The vibrations among different rotor speeds have similar varying tendency to a certain extent.

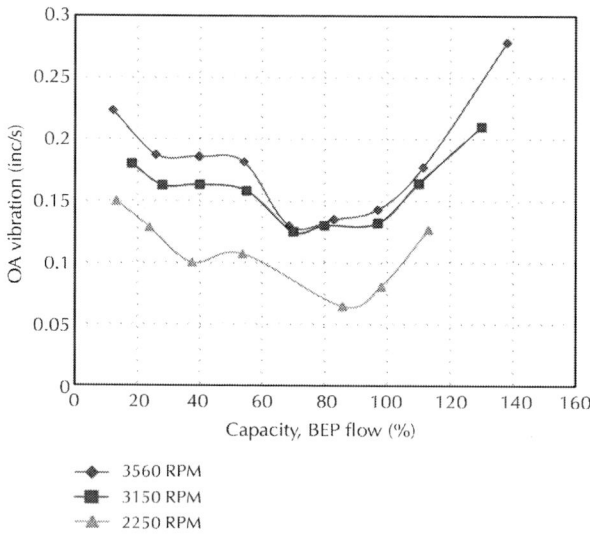

Figure 3: Test vibration level dates in different pump obtained by Stavale.

However, in centrifugal pumps, the vibration spectrum usually contains only the fundamental shaft frequency, the blade passage frequency, and one or two harmonics. All these characteristic frequencies are directly decided by the rotor speed. Therefore, potential dangers of resonance exist, and when resonance occurs at the characteristic frequency close to the natural frequency, the similar varying tendencies are gone. Thus, in practice, lockout speed range is recommended.

As all forces are proportional to the product of pressure × area, $F \sim n^2 \times d^4$ also applies.

For a given pump, the hydraulic excitation forces increase with the square of the tip speed and the density, that is, $\rho \times u^2$. The bearing housing and shaft vibrations of the pump would also increase with $\rho \times u^2$ if it were not for resonance phenomena.

In Stavale's work [10], the overall vibration was measured for a variable-speed test and compared with a constant-speed system with throttle valve, as shown in Figure 4. During the variable-speed test, the system was fixed, that is, there was no throttling of the control valve to initiate changes in the flow. Prior to the variable-speed test, the control valve was opened wide, and the backpressure valve was adjusted to obtain maximum flow at maximum speed; the backpressure was approximately zero.

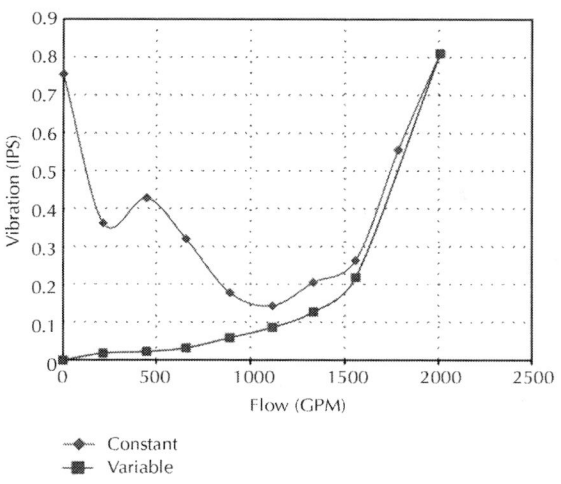

Figure 4: Vibration levels for different operating models of the same pump.

Figure 4 shows that in the course of the variable-speed operation, the pump can operate by following the same operating condition line for the approximately zero backpressure. The vibration level of the square of the flow under the same operating condition line can be clearly seen, so the vibration also follows the same law for the resonance phenomena of a given pump $V_e \sim n^2$, expressed as

$$\frac{V_e(n_1)}{V_e(n_2)} = \frac{n_1^2}{n_2^2},$$

(2.2)

where n_1 and n_2 represent two different operating pump speeds.

MAINTENANCE COST MODEL IN PUMP OPERATION

Maintenance costs cannot be easily estimated, however, the wear of pumps is mainly caused by frequent switching them on and off. Formally, a pump switch is defined as turning on a pump which was previously off. Therefore, minimizing the number of pump switches will result in minimization of maintenance costs. As a result, the traditional method assumes that it increases as the number of "pump switched" increases, and its model can be built according to Lansey and Awumah [5] and Vladimir et al. [14].

In many applications, the cost of unscheduled maintenance which was not considered in former model may be the largest contributor to operation cost in process plants, and although with the mainly aspect of the wear in switch course, the operation course should not been ignored.

The life of the mechanical seal is directly related to shaft movement. Vibration can cause carbon face chipping and seal face opening. Drive lugs will wear, and metal bellows seals will fatigue. In some instances, the shaft movement can cause the rotating seal components to contact the inside of the stuffing box, or some other stationary object, causing the seal faces to open and allowing solids to penetrate between the lapped faces. Vibration is also a major cause of set screws becoming loose and slipping on the shaft, causing the lapped seal faces to open. The vibration would also cause denting of the bearing races for no design for the bearings to handle both a radial and axial load.

Critical dimensions and tolerances such as wear ring clearance and impeller setting will be affected by vibration.

Bearing seals are very sensitive to shaft radial movement. Shaft damage will increase and the seals will fail prematurely. Labyrinth seals operate with a very close tolerance. Excessive movement can damage these tolerances also.

Vibration can affect the reliability and life of the equipment and is often assumed to be the direct as direct sign of the reliability and health of the machine.

As vibration can affect the reliability and life of the equipment, indicators must be set up to consider vibration data. Maintenance cost is further assumed to increase as the reliability indicators reduce. In addition, maintenance cost can be significantly reduced with operation in a high-reliability mode, thus reducing wear and reliability incidents.

Indicators of Operation Reliability

Vibration data can be normalized based on the following equation:

$$R = \left(1 - \left(\frac{V}{V_{max}}\right)\right) + C,$$

(3.1)

where V is the vibration level at a certain point, V_{max} is the maximum date value, C is a constant added to set the peak value of R (equal to one) equal to V_{min}/V_{max}, and R is the relative reliability indicator.

The R value in (3.1) is a relative reliability number between zero and one. A zero value does not necessarily indicate zero reliability but is rather intended to discourage running the pump at these conditions. Similarly, a value of one does not indicate infinite reliability but is intended to be a relative indicator of the best operating conditions for a given pump. As the mechanical design of a pump can also affect reliability, these values should not be used to compare pumps of different designs or manufacturers. It is intended to compare the effect of alternative operating conditions on reliability.

Maintenance Cost Model during Operation Based on Reliability Indicators

Single-Pump Model

For a fixed-speed pump, the mathematical model can be expressed as a set of static relationships between the vibration level and flow rate. The pump vibration level is referred to as vibration velocity or vibration displacement, expressed as RMS or peak-to-peak values. The vibration level data are divided by the lowest of all the flows for simplification and normalization. The flow rate is also referred to as capability. Then, a single model can be expressed as a cubic or fourth-order polynomial equation for the bathtub-shaped curve:

$$V = \overline{v}_0 + \overline{v}_1 Q + \overline{v}_2 Q^2 + \overline{v}_3 Q^3, \tag{3.2}$$

where V is the normalized vibration level at a certain point, Q is the flow rate, and system parameters v_i are determined by specific pump vibration characteristics and can be identified by test data.

With regard to the variable-speed pump, the relationship between (3.2) and its parameters is motor-speed dependent. The affinity law in pump theory states the following:

$$\frac{Q(n_1)}{Q(n_2)} = \frac{n_1}{n_2}, \qquad \frac{H(n_1)}{H(n_2)} = \frac{n_1^2}{n_2^2}, \qquad \frac{P(n_1)}{P(n_2)} = \frac{n_1^3}{n_2^3}, \tag{3.3}$$

where n_1 and n_2 represent two different operating pump speeds.

Assuming that the pump model (3.2) for a VSD at a special speed n_0 is obtained and its indicators are v_0, v_1, v_2, and v_3, the pump model of the considered VSD for a given limited speed n has the property defined by (2.2) and (3.2), that is,

$$\frac{V(n)}{V(n_0)} = \frac{V(n)}{v_0 + v_1 Q(n_0) + v_2 Q(n_0)^2 + v_3 Q(n_0)^3} = \frac{n^2}{n_0^2},$$

$$V(n) = \frac{n^2}{n_0^2}\left[v_0 + v_1 Q(n_0) + v_2 Q(n_0)^2 + v_3 Q(n_0)^3\right].$$
(3.4)

Then, according to (3.3), with k=n/n$_0$, this part can be transformed into

$$V(n) = \frac{n^2}{n_0^2}\left[v_0 + v_1\left(Q(n) * \frac{n_0}{n_1}\right) + v_2\left(Q(n) * \frac{n_0}{n_1}\right)^2 + v_3\left(Q(n) * \frac{n_0}{n_1}\right)^3\right]$$

$$= k^2 v_0 + k v_1 Q(n) + v_2 Q^2(n) + k^{-1} v_3 Q^3(n).$$
(3.5)

Subsequently, according to (3.1), the operation reliability of the signal pump can be expressed as

$$R_{\text{sig}} = \left(1 - \left(\frac{V(n,Q)}{V_{\text{max}}}\right)\right) + \frac{V_{\text{min}}}{V_{\text{max}}}.$$
(3.6)

R_{sig} is a function of Q (flow rate) and n (rotation speed). It is directly related to the operating conditions and reflects the relationship between the maintenance cost and the operating conditions. The higher R_{sig} is, the lower is the maintenance cost in the course of operation.

MultiPump Model

The general multiple pump systems consist of some special characteristics for operating maintenance cost analysis.

- The model should only reflect the pumps that are in operation. Pumps that are not running should not be considered.
- As maintenance cost is not the same for different pumps, the model should consider the difference.
- The relative normalization number is also between zero and one, similar to the single-pump model.

Considering the above-mentioned factors, the following methods are adopted.

- A swift variable vector is added to represent the operating condition of the pump group. Then, only the pump in operation is considered; hence, the program (2.1) can be solved.
- As the cost of pump is usually related to the installed capacity, assumption is made that the bigger the installed capacity is, the higher is the maintenance cost. Thus, a weight constructed with the ratio of single installed capacity to total installed capacity is adopted to represent the different maintenance costs.

From the above discussion, conclusion is made that the maintenance cost model for multipump system with k pumps can be expressed as follows:

$$R = \sum_{i=1}^{k} \frac{w_i \varphi_i R_{\text{sig}i}}{w_i \varphi_i}, \qquad (3.7)$$

Where

$$w_i \in \{0, 1\}, \quad i = 1, 2 \ldots k. \qquad (3.8)$$

This equation is the swift variable vector. "1" shows that the pump is working, and "0" indicates that the pump is off:

$$\varphi_i = \frac{N_{di}}{N_{dT}}, \quad i = 1, 2 \ldots k. \qquad (3.9)$$

N_{dT} is the total installed capacity of the multipump system, and N_{di} is single installed capacity of the ith pump.

Modeling Method of the Maintenance Cost without Test Data

As environment and reliability requirements increase, the pumps are built and tested to standards and specifications that define the maximum allowable vibration amplitudes (process pumps and almost all large pumps) in many applications. However, unlike the pump characteristics, some installations still do not have such vibration data because no mandatory standard is required. Thus, the maintenance cost model may not work. Hence, other modeling methods must be developed to deal with this situation.

Fortunately, although specific vibration data may not be known, the allowable vibration amplitudes defined in standards, such as API 610 [15], are clear. The standards define different vibration limits for a "preferred operation range" and an "allowed operation range." In the preferred operation range, the maximum increment of vibration is less than 10%. On the other hand, the maximum increment of vibration is less than 30% for the allowed operation range. In a certain pumping system, if the preferred and the allowed operation ranges are known, assumption can be made that all systems meet the standard. Therefore, a maintenance cost model can be developed.

Even if the preferred and allowed operation ranges are unknown, API 610 still provides an idealized flow versus vibration plot [11], as shown in Figure 5. Maintenance cost model can also be developed based on the assumption that the pump follows this law.

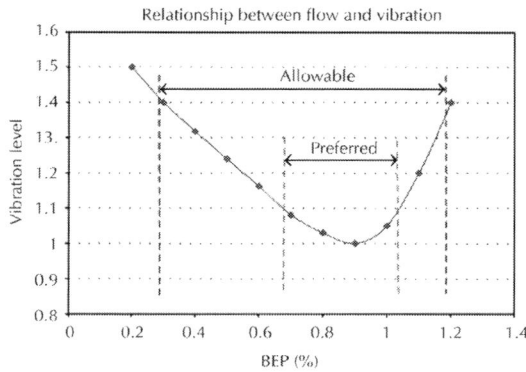

Figure 5: Idealized flow versus vibration plot recommended by API 610.

OPTIMIZATION MODEL

Generally, the optimal policy should result in the lowest total operating cost for a given set of boundary and system constraints. Thus, objective function and constraints are needed for optimal scheduling model.

Objective Function

In a typical pumping system, the operation cost mostly comprises energy-consumption cost and maintenance cost. Thus, the objective function also comprises these two factors [16–19].

Energy Cost

Pump scheduling is used for dealing with the following two situations.

One is the selection process on which available pumps are to be used and for what period of time (e.g., a day) the pumps should operate.

The other is referred to as (real-time) control problem. The optimal strategy is concerned with which available pumps must be operated and when they should be operated according to fluctuations in demands and/or operating conditions.

For the first situation [14], the objective is the determination of energy consumption of all pumps in the pumping station during the optimization period. The charging structure used by the electric utility is the factor that must be considered in analyzing electric energy cost. Then, the objective function is mathematically expressed as

$$E = \sum_{i=1}^{I}\sum_{t=1}^{T}(ER_{it})CQ_{it}(HS_{it}),$$

(4.1)

where E is the total energy cost to be minimized, I is number of pump systems, and T is number of time intervals that constitute the operating horizon. ER_{it} is the electrical rating of pump i during time period t, C is the conversion coefficient, and Q_{it} and HS_{it} are the discharge and pressure head, respectively, of pump i during time period t.

For the second situation [16], the real scheme is to consider the total input electrical power as the objective function. This method is much closer to the real situation. However, the calculation time will be long, which may not be suitable for real-time control. Thus, the modeling method commonly used is by considering the shaft horsepower of the pumps as objective function because the energy consumption of the motor and the inverter is very much less than that of the pump. Then, the objective function can be mathematically expressed as

$$f1 = \sum_{i=1}^{I} w_i P_i(Q_i, k_i),$$

(4.2)

where f1 is the total shaft horsepower to be minimized, I is number of pumps, and Q_i and k_i are the discharge and pressure head, respectively, of pump i under certain operating conditions. Pi is the shaft horsepower of pump i under this condition, which can be expressed as (3.9) according to the characteristic of the pump and the affinity law, and p_i is theconversion coefficient:

$$P(Q, k) = p_0 k^3 + p_1 k^2 Q + p_2 k Q^2 + p_3 Q^3.$$

(4.3)

Pump Maintenance Cost

The pump maintenance cost can be as important as the electric energy cost or even more relevant; however, this cost cannot be quantified easily, but it can be described through other correlative factors indirectly. Thus, the number of "pump switched" and the operation relative reliability indicators are chosen to describe this factor.

The Number of "Pump Switched"

Due to the difficulty in starting-up the pumps and the significant increase in frequency during their switching, the maintenance cost increases as the number of "pump switches" increases, which can be expressed as follows when real control is used for the pump schedule:

$$d_H(w, w') = \sum_{i=1}^{n} |w_i - w_i'|,$$

(4.4)

where I is the number of pumps, w_i is the operating status of pump i for the next step, w_i' is the present operating status, w_i and w_i' are both switch variables, 1 indicates that the pump is in operation, and 0 shows that the pump is off. When the pump schedule is used for this scheme for a period of time, the objective function becomes the total switch number at this time, expressed as

$$N_s = \sum_{t=1}^{T} \sum_{i=1}^{n} |w_i^t - w_i^{t-1}|.$$

(4.5)

Operation Relative Reliability Indicators

R in (3.6) and (3.7) is a relative reliability number between zero and one. As R approaches one, the higher the reliability is, the lower is the pump wear and the longer is the MTBR; subsequently, maintenance cost will be reduced. For the real-time control, (3.7) can be used directly to express the scheme in a period of time. The summation of the time periods can be used as the objective function.

Constraints

Water-Supply Demand Constraints

For the pump system to meet some of the water-supply requirements [17], the desired watersupply index, which is known for the pump system, can be expressed as (H_{ST}, Q_e), expressed in a mathematic model.

For parallel-connected pump systems,

$$Q_e = \sum_{i=1}^{n} Q_i,$$

$$H_{ST} = H_i. \tag{4.6}$$

For series-connected pump systems,

$$Q_e = Q_i, \tag{4.7}$$

$$H_{ST} = \sum_{i=1}^{n} H_i. \tag{4.8}$$

Operation Constraints

Obviously, a pump should be selected so that it operates predominantly close to the BEP in the so-called "preferred operation range." This mode of operation is apt to produce the lowest energy and maintenance costs and minimize the risk of system problems. However, off-design operation for limited periods cannot be avoided. Rules are needed to define the allowable ranges and modes of operation to reduce the risk of damage and excessive wear. To this effect, limits must be defined for continuous and for short-term operations at maximum and minimum flow [16].

The range of preferred continuous operation can be defined, for instance, by the requirement that the efficiency must not fall below 80% to 85% of the maximum efficiency of the pump, and allowable ranges can be defined so that the efficiency must not fall below 70%.

This constraint can be expressed as

$$\min(Q_{POPi}) \leq Q_i \leq \max(Q_{POPi}), \tag{4.9}$$

$$\min(Q_{AOPi}) \le Q_i \le \max(Q_{AOPi}). \tag{4.10}$$

For optimal operation, the constraint should be in POP, as shown in (4.9). However, for some situations, the optimization model may have no solution. In this case, the constraint is adjusted to AOP; then, (4.10) can be applied.

At the same time, when the pump is operated by VSD, the speed regulation range should also be constrained. The operation efficiency will decrease because of a very wide speed range, and reliability will reduce. Therefore, the factors for operation should be considered. The speed regulation range is constrained as $[k_{min}\ 1]$ k_{min} is determined by several factors, such as the rotation speed at which the pump is no longer able to maintain discharge against the static head from the demand side, the rotation speed to avoid system resonance, and the rotation speed to ensure that the pump operating economically.

As the pump operates under speed-control model, the range of the continuous and short-term operations will constantly change.

According to the affinity law in pump theory, l_1 and l_2 _Figure 6_ are set as similar lines, and A, B, C, and D are the boundaries of the operation zone; then, l_1 and l_2 can be expressed as

$$H_{l1} = \frac{H_A}{Q_A}Q_{l1}^2, \qquad H_{l2} = \frac{H_B}{Q_B}Q_{l2}^2. \tag{4.11}$$

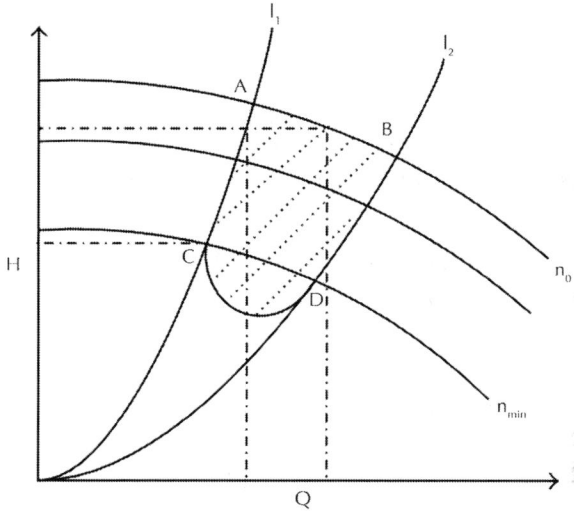

Figure 6: Efficient operation area for the variable-speed pump.

If head H_e is needed for this water supply system, then the boundary could be changed:

$$Q_{min} = \begin{cases} Q_A\sqrt{\dfrac{H_e}{H_A}} & H_e \geq H_C, \\ Q_c & H_e < H_C, \end{cases}$$

$$Q_{max} = \begin{cases} Q_B\sqrt{\dfrac{H_e}{H_B}} & H_e < H_B, \\ Q_B & H_e \geq H_B. \end{cases}$$

(4.12)

APPLICATION

Profiles of the Pump Station

A sample model is a circulating water pumping station, one of the most important facilities in an alumina plant used for the mother liquor evaporation process. It uses almost 17% of the electricity consumption in the plant, and the reliability of this system is very important for the whole plant.

There are five pumps of single- and double-stage suction in the pumping system, and the pump model is shown in Table 1.

Table 1: The configuration and the parameters of the pump

Serial number	Pump model	Qden (m3/h)	Hden m	Qmin m3/h	Qmax m3/h	N (RPM)
1#	14SH-9B	1425	58	855	1853	1450
2#	20SA-10	2850	58	1710	3848	960
3#	20SA-10	2850	58	1710	3848	960
4#	20SA-10	2850	58	1710	3848	960
5#	20SA-10	2850	58	1710	3848	960

Variable-speed control is adopted for $1^\#$ and $2^\#$, and the minimum speed regulation ranges (k_{min}) are 0.7 and 0.75 in this system.

The operation characteristics of these pumps are shown in Table 2.

Table 2: The model parameters of the pump

$H=Hx=SQ2$				
Serial number	Hxi	Si		
1#	$71.17\ k_1^2$	7.488e-6		
2#	$70.39\ k_2^2$	1.780e-6		
3#	70.39	1.780e-6		
4#	70.39	1.780e-6		
5#	70.39	1.780e-6		
$P=P0+P1Q+P2Q2+P3Q3$				
Serial number	P0	P1	P2	P3
1#	$146.4\ k_1^3$	$0.05\ k_1^2$	$4.4e\ k_1$	−1.44e-8
2#	$230.5\ k_2^3$	$0.1025\ k_2^2$	$5.826e\ k_2$	−2.1e-9

3#	230.5	0.1025	5.826e–6	–2.1e-9
4#	230.5	0.1025	5.826e–6	–2.1e-9
5#	230.5	0.1025	5.826e–6	–2.1e-9
V=V0+V1Q+V2Q2+V3Q3				
Serial number	V0	V1	V2	V3
1#	$7.696\text{e}{-}10\,k_1^2$	$-1.57\text{e}{-}6\,k_1$	3.84e–4	$1.483\,k_1^{-1}$
2#	$9.62\text{e}{-}11\,k_2^2$	$-3.936\text{e}{-}7\,k_2$	1.92e–4	$1.483\,k_2^{-1}$
3#	9.62e–11	–3.936e–7	1.92e–4	1.483
4#	9.62e–11	–3.936e–7	1.92e–4	1.483
5#	9.62e–11	–3.936e–7	1.92e–4	1.483

For this pumping system, the basic demand characteristic obeys (4.10) and can be calculated based on process requirements and pipeline characteristics.

$$H_{\mathrm{dem}} = 42 + 3.51 \times 10^{-7}Q^2. \tag{5.1}$$

The whole flow date is shown in Figure 7, and these values are obtained from the water demand curve based on historical data.

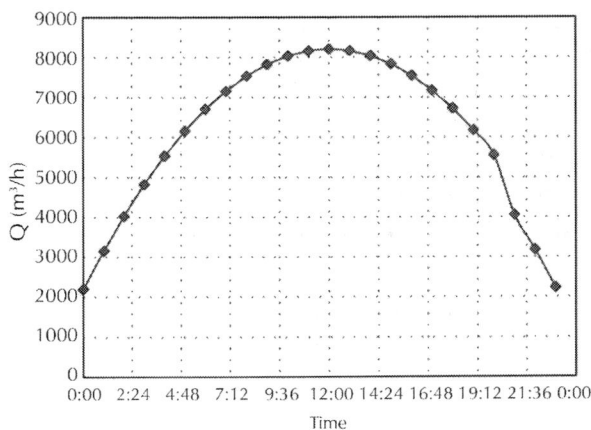

Figure 7: Daily variation curve of the entire flow demand.

Optimization Pump Scheduling and the Result

A one-day optimization result was made. Assumption was made that the shortest period for each combination of pumps is 1 h, that is, a pump can be switched off/on after having been active or inactive for at least 1 h.

Based on the flow demand in Figure 7 and the head demand in (4.10), the desired objective value is shown in Figure 8.

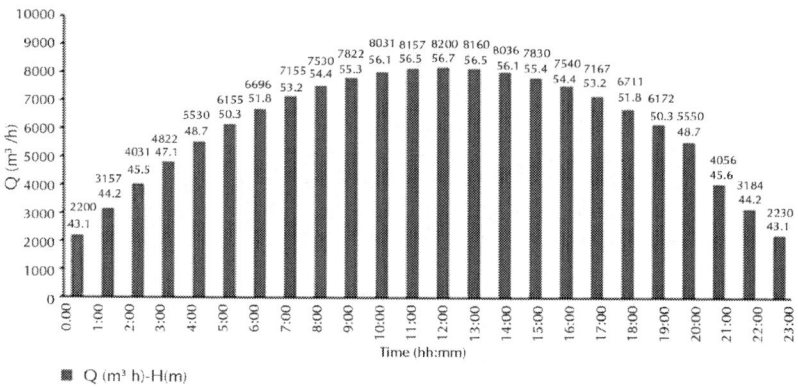

Figure 8: Daily supply index.

Genetic algorithm was selected as the optimization method because of its suitable characteristics for adaptability to complex optimization problem [18, 19].

In this optimization problem, the single objective approach developed by Mackle was adopted because of its simplicity. The fitness function consists of the energy consumption cost and penalties for violation of the constraints of the system. All these factors were linearly weighted.

Then, the optimal result is shown in Figures 9, 10, and 11. A represents the result when only the minimum energy cost is considered, and B represents the result when both maintenance and energy costs are considered.

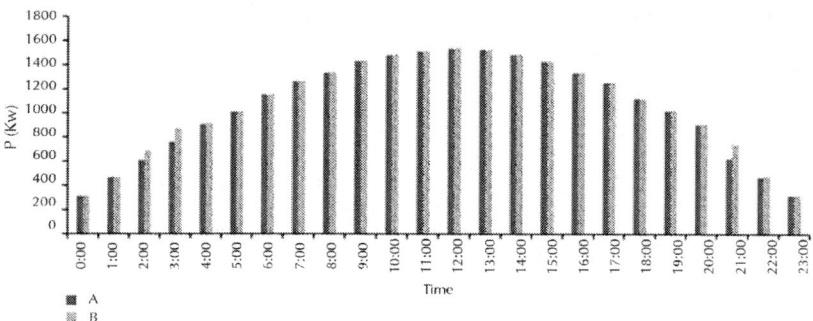

Figure 9: Input power of the different models.

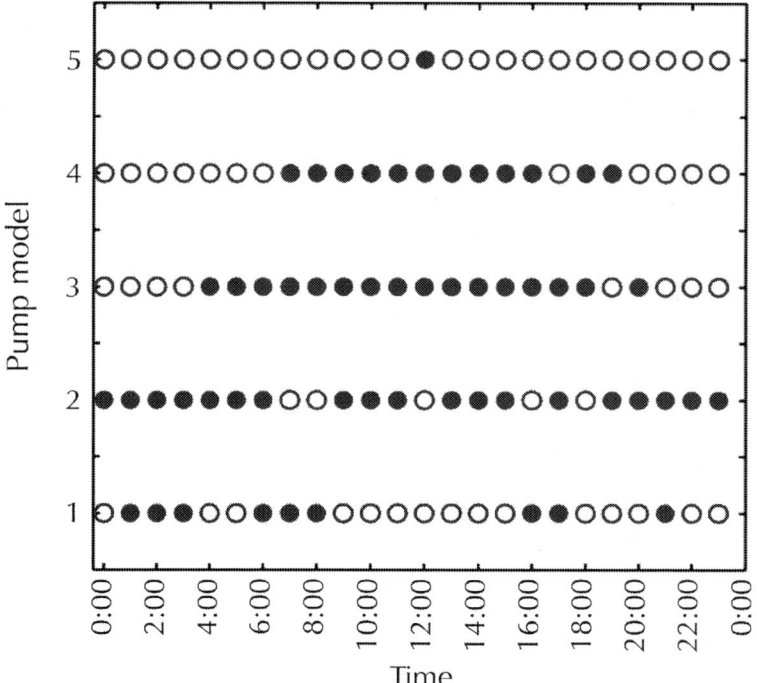

(a)

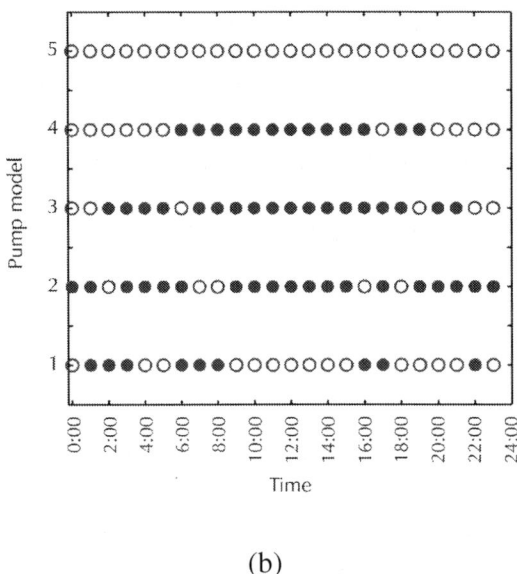

(b)

Figure 10: (a) Operation state of the form model. (b) Operation state of the new improved model.

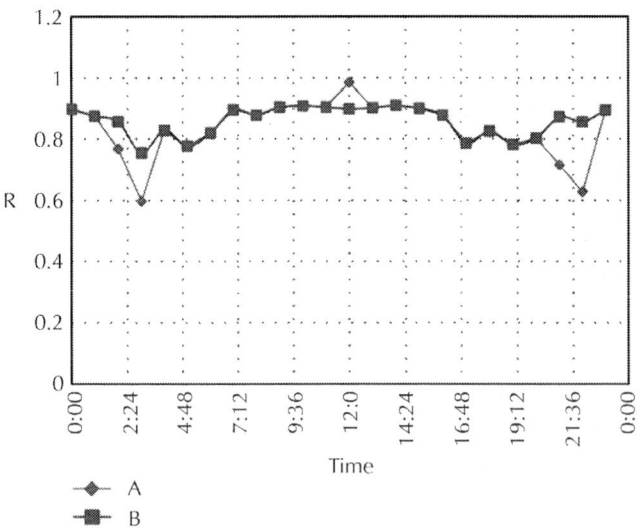

Figure 11: Reliability factor number of the different models.

Figure 9 shows the result of the shaft power in the two different optimization objectives under the same water supply index. The result considered that the energy consumption was smaller, and the multiobjects are almost equal to the whole, whereas in some cases, they were smaller. This condition might be because the high efficiency point is the most reliable operation point in most situations; however, in some cases, the general shaft power may be lower. For a single pump, operating in the high vibration point can cause the whole system to operate in unreliable conditions.

Thus, from energy-saving perspective, considering only energy cost as the main objective may be even better. However, when the switch number and reliability factor are considered in the optimization course, pump control are easy and simple using the least number of switches, and reliability will be enhanced when the operation point is close to the low vibration point as much as possible. Then, the pump system will be operating under high reliability condition, the maintenance cost will be reduced, and the service life will be extended. Subsequently, the whole LCC will be decreased by this kind of optimization scheduling model.

CONCLUSIONS

In the present work, a new multiobjective approach, which takes into account operation reliability and maintenance cost incurred in operation, is presented.

The main conclusions from the current study are the following.

- Operation cost reduction and operation reliability enhancement are the focus of operation control for pump systems. Operation reliability for running pumps, which is a key to decrease unscheduled maintenance costs caused by the reliability incidents, and the wear cost during the operation should be considered.

- Vibration can affect the reliability and life of the equipment. An index reflecting the operation reliability and the wear degree in operation course could be considered as the normalization of vibration level.

- The idealized flow versus vibration plot should take a distinct bathtub shape. For this bathtub shape, a narrow sweet spot (80% to 100% BEP) exists that can be used to obtain low vibration levels, and the vibration also follows a similar law with the square of the rotation speed if it were not for resonance phenomena with a given pump.

- The maintenance cost which is not considered by traditional model concluding the unscheduled maintenance cost and the wear cost during the operation can be modeled as a function of the pump capacity and rotation speed based on the vibration characteristics. This function is then added to the traditional optimal scheduling model to create a new optimal scheduling model.

- Compared with the traditional method, the result of the new optimal model changes the result produced by the traditional one. It improves the operating conditions of the pump and enables the pump to run operate low vibration level. Therefore, maintenance cost can be reduced and the operation reliability can be enhanced to a certain degree.

ACKNOWLEDGMENTS

This work is supported by National Outstanding Young Scientists Founds of China (Grant no. 50825902), Jiangsu Provincial Innovative Scholars "Climbing" Project of China (Grant no. BK 2009006), the National Natural Science Foundation of China (Grant no. 50979034), and Priority Academic Program Development of Jiangsu Higher Education Institutions.

REFERENCES

1. Hydraulic Institute, Pump Life Cycle Costs: A Guide to LCC Analysis for Pumping Systems, Parsippany, NJ, USA, 2001.

2. Z.-H. Wang, G.-H. Geng, and S.-K. Song, "Discussion on energy conservation of pump," China Foreign Energy, vol. 11, no. 5, pp. 73–76, 2006 (Chinese).

3.   L. W. Mays, Water Distribution Systems Handbook, McGraw-Hill, New York, NY, USA, 2000.

4.   L. E. Ormsbee and K. E. Lansey, "Optimal control of water-supply pumping systems," Journal of Water Resources Planning & Management, vol. 120, no. 2, pp. 237–252, 1994. ··

5.   K. E. Lansey and K. Awumah, "Optimal pump operations considering pump switches," Journal of Water Resources Planning and Management, vol. 120, no. 1, pp. 17–35, 1994. · ·

6.   D. C. White, "Improve your project›s prospects," Chemical Processing, vol. 67, no. 10, pp. 33–39, 2004. ·

7.   H. P. Bloch and F. K. Geitner, An Introduction Machinery Reliability Assessment, Gulf Publishing, Houston, Tex, USA, 1994.

8.   A. R. Budris, R. B. Erickson, F. H. Kludt, and C. Small, "Consider hydraulic factors to reduce pump downtime," Chemical Engineering, vol. 109, no. 1, pp. 54–60, 2002.

9.   R. B. Erickson, E. P. Sabini, and A. E. Stavale, Hydraulic Selection to Minimize the Unscheduled Maintenance Portion of Life Cycle Cost, Pump Users International Forum, Karlsruhe, Germany, 2000.

10.   A. E. Stavale, "Reducing reliability incidents and improving meantime between repair," in Proceedings of the 24th Interenational Pump Users Symposium, pp. 1–10, College Station, Tex, USA, 2008.

11.   X. Robert and P. E. Perez, "Operating Centrifugal Pumps Off-design- pumps & systems 20 suggestions for a new analysis method operating centrifugal pumps," April 2005, http://www.pump-zone.com/articles/2.pdf.

12.   Y.-Y. Ni, 3-D unsteady numerical simulation and fluid-induced vibration for centrifugal pumps, Ph.D. thesis, Jiangsu University, ZhenJiang, China, 2008.

13.   Y.-Y. Ni, S.-Qi Yuan, Z.-Y. Pan, et al., "Diagnosing the running condition of pump by its vibration character," Drainage and Irrigation Machinery, vol. 25, no. 2, pp. 49–52, 2007 (Chinese).

14.   C. Vladimir, M. Heiliö, N. Kreji , and M. Nedeljkov, "Mathematical model for efficient water flow management," Nonlinear Analysis, vol. 11, no. 3, pp. 1600–1612, 2010. · ·

15. API Standard 610, Centrifugal Pumps for Petroleum Heavy Duty Chemical and Gas Industry Services, Petrochemical and Natural Gas Industries, 9th edition, 2003.

16. C. Zhang, H. Li, M. Zhong, and J. Cheng, "The modelling and optimal scheduling for pressure and flow varying parallel-connected pump systems," Dynamics of Continuous, Discrete and Impulsive Systems B, vol. 11, no. 6, pp. 757–770, 2004. ·

17. T. C. Yu, T. Q. Zhang, and X. Li, "Optimal operation of water supply systems with tanks based on genetic algorithm," Journal of Zhejiang University, vol. 6, no. 8, pp. 886–893, 2005. · ·

18. B. Barán, C. Von Lücken, and A. Sotelo, "Multi-objective pump scheduling optimisation using evolutionary strategies," Advances in Engineering Software, vol. 36, no. 1, pp. 39–47, 2005. · ·

19. A. S. Dragan, A. W. Godfrey, and S. Martin, "Multiobjective genetic algorithms for pump scheduling in water supply," Lecture Notes in Computer Science: Evolutionary Computing, vol. 1305, pp. 227–223, 1997. ·

Citations

CHAPTER 1

K. Tennokese, T. Kõiv, A. Mikola and V. Vares, "The Application of the Ground Source and Air-to-Water Heat Pumps in Cold Climate Areas," Smart Grid and Renewable Energy, Vol. 4 No. 7, 2013, pp. 473-481. doi:10.4236/sgre.2013.47054.

CHAPTER 2

S. Anish, N. Sitaram and H. Kim, "Study of Secondary Flow Modifications at Impeller Exit of a Centrifugal Compressor," Open Journal of Fluid Dynamics, Vol. 2 No. 4A, 2012, pp. 248-256. doi: 10.4236/ojfd.2012.24A029.

CHAPTER 3

Yuliang Zhang, Zuchao Zhu, Yingzi Jin, Baoling Cui. Transient Hydraulic Performance and Numerical Simulation of a Centrifugal Pump with an Open Impeller during Shutting Down Doi.org/10.4236/ojfd.2012.24A044.

CHAPTER 4

Massinissa Djerroud, Guyh Dituba Ngoma, and Walid Ghie, "Numerical Identification of Key Design Parameters Enhancing the Centrifugal Pump Performance: Impeller, Impeller-Volute, and Impeller-Diffuser," ISRN Mechanical Engineering, vol. 2011, Article ID 794341, 16 pages, 2011. doi:10.5402/2011/794341.

CHAPTER 5

Qihua Zhang, Weidong Shi, Yan Xu, et al., "A New Proposed Return Guide Vane for Compact Multistage Centrifugal Pumps," International Journal of Rotating Machinery, vol. 2013, Article ID 683713, 11 pages, 2013. doi:10.1155/2013/683713.

CHAPTER 6

Zhi-jian Wang, Jian-she Zheng, Lu-lu Li, and Shuai Luo, "Research on Three-Dimensional Unsteady Turbulent Flow in Multistage Centrifugal Pump and Performance Prediction Based on CFD," Mathematical Problems in Engineering, vol. 2013, Article ID 589161, 7 pages, 2013. doi:10.1155/2013/589161.

CHAPTER 7

Yu Zhang, Sanbao Hu, Yunqing Zhang, and Liping Chen, "Optimization and Analysis of Centrifugal Pump considering Fluid-Structure Interaction," The Scientific World Journal, vol. 2014, Article ID 131802, 9 pages, 2014. doi:10.1155/2014/131802.

CHAPTER 8

Julian Fasano, Eric E. Janz, and Kevin Myers, "Design Mixers to Minimize Effects of Erosion and Corrosion Erosion," International Journal of Chemical Engineering, vol. 2012, Article ID 171838, 8 pages, 2012. doi:10.1155/2012/171838.

CHAPTER 9

Hong Li, Haifei Zhuang, and Weihao Geng, "Design of a Turbulence Generator of Medium Consistency Pulp Pumps," International Journal of Rotating Machinery, vol. 2012, Article ID 413674, 7 pages, 2012. doi:10.1155/2012/413674.

CHAPTER 10

Yin Luo, Shouqi Yuan, Yue Tang, Jianping Yuan, and Jinfeng Zhang, "Modeling Optimal Scheduling for Pumping System to Minimize Operation Cost and Enhance Operation Reliability," Journal of Applied Mathematics, vol. 2012, Article ID 370502, 19 pages, 2012. doi:10.1155/2012/370502.

Index